ELECTRONIC MOTOR AND WELDER CONTROLS

Electronic Motor and Welder Controls

BY

GEORGE M. CHUTE

Application Engineer, General Electric Company, Detroit

FIRST EDITION

McGRAW-HILL BOOK COMPANY, INC.
NEW YORK TORONTO LONDON
1951

ELECTRONIC MOTOR AND WELDER CONTROLS

Copyright, 1951, by the McGraw-Hill Book Company, Inc. Printed in the United States of America. All rights reserved. This book, or parts thereof, may not be reproduced in any form without permission of the publishers.

To
JO, MARJORIE AND GEORGE, MARION AND BOB

PREFACE

Recent industrial uses of electronic equipment continue to show that two major groups of tube-operated equipment are employed in resistance welding and for the control of motors in a wide variety of industrial processes. These equipments of advanced design, now appearing in industrial plants throughout the country, warrant continued study by those responsible for their operation and maintenance.

In describing recent types of welding control, the author adds to the information given in "Electronic Control of Resistance Welding." There is little repetition of earlier material. Combinations of novel basic electron-tube circuits open new fields of interest, as in the all-tube sequence timers and the frequency-changer type of welder control. In describing these more complex recent circuits, it is assumed that the reader is familiar with simpler electronic controls, but there is frequent reference to the location of background material as needed.

While some of the general-purpose types of electronic motor control have been covered in "Electronics in Industry," improved electrical circuits now are presented, with a much greater variety of closed-cycle motor-control systems. The stable operation of such motor systems is considered.

The author wishes to acknowledge the assistance of the Electronic and Regulator Control Divisions of the General Electric Company, especially that of J. T. Bailey, H. L. Palmer and their coworkers. Photographs have been supplied by General Electric.

GEORGE M. CHUTE

DETROIT, MICHIGAN
January, 1951

CONTENTS

PREFACE vii

Part I
RESISTANCE WELDING CONTROLS

1. BASIC WELDING CONTROLS 3
   Resistance welding—Ignitron contactor—NEMA 1A timer (CR7503–F220)—Phase-shift heat control (CR7503–D175)—Load rating of Ignitrons—Flip-flop control for larger loads (CR7503–G119).

2. G–E SEQUENCE WELD TIMERS 20
   NEMA 3B and 5B timers—Pulsation welding—Pulsation timer (CR7503–F180)—Combination timer and contactor (CR7503–L102A8)—Dual weld times.

3. MORE SEQUENCE TIMERS 37
   Weltronic 3B timer (Model 75SE2)—Taylor-Winfield N3 combination—Westinghouse N2 combination.

4. G–E HIGH-SPEED SEQUENCE COMBINATIONS 55
   NEMA N2 combination (CR7503–L102S15)—Thyratron action with resistance or inductive load—"Inductive hangover"—Automotive timer (CR7503–L127)—Double squeeze time.

5. SYNCHRONOUS TIMING 74
   Prevention of transients—Leading-tube-trailing-tube action—Synchronous control for small welders (CR7503–A140).

6. G–E SYNCHRONOUS COMBINATIONS 83
   Spot-welding S2H combination (CR7503–A151)—Pulsation-welding S3H combination (CR7503–A152).

CONTENTS

7. SLOPE CONTROL, TEMPER WELDING AND FORGE TIMING   100
   Slope control (CR7503–D202)—Synchronous combination for temper welding (CR7503–A160)—Synchronous combination with forge delay (CR7503–A164).

8. G–E SEAM-WELDING CONTROLS   114
   Synchronous seam-welding control (CR7503–B120)—Spot, seam, pulsation-welding combination (CR7503–C121).

9. WESTINGHOUSE SYNCHRONOUS COMBINATIONS . . .   129
   Spot-welding S2H combination—Pulsation-welding S3H combination—Seam-welding control.

10. WELDING WITH LIMITED POWER SUPPLY   144
    Low-voltage control with series capacitors—High-voltage synchronous control (CR7503–A148).

11. THREE-PHASE WELDING CONTROLS   155
    Multi-transformer welding machines—Lower frequency for welding—Frequency-changer spot-welding control (CR7503–M101)—Inversion for faster current decay.

Part II
MOTOR CONTROLS

12. POSITION CONTROLS   179
    Electronic positioning control (CR7510–A102)—Closed-cycle systems—Selsyn control of positioning—Two-point side-register control (CR7505–S118).

13. SPEED CONTROL OF D-C MOTORS   191
    Basic d-c motor behavior—Photoelectric loop control (CR7505–T102)—Phase-shifting by saturable reactor.

14. PHOTOELECTRIC REGISTER CONTROLS   203
    One-way cutoff control (CR7515–W201)—Side-register control (CR7515–S127)—All-tube side-register control (CR7515–S120)—Two-way cutoff-register control (CR7515–W108).

CONTENTS

15. THY-MO-TROL—AUTOMATIC TUBE CONTROL OF D-C MOTORS 226

    Half-wave supply to d-c motor (CR7507–F170)—Full-wave Thy-mo-trol system (CR7507–G270)—Speed control by solenoid movement (CR7507–N115)—By selsyn (CR7507–N103).

16. MOTOR-GENERATOR CONTROL OF MOTOR SPEED . . 261

    Electronic speed variators—timed-acceleration variator (CR7513–E101)—Variator with current-limit control (CR7513–E118).

17. THE ELECTRONIC AMPLIDYNE 278

    Electronic amplidyne (5AM79AB334)—Constant motor speed with current limit—Tachometer feedback—Constant motor current with speed limit.

18. STABILITY OF A CLOSED-CYCLE SYSTEM 289

    Antihunt circuits—The Bode diagram—Phase-lag, phase-lead, notch networks.

19. THE PHOTOELECTRIC TRACER 303

    Electronic tracer (CR7512–B100)—Line-follower action.

20. RUBBER CALENDER CONTROL 310

    Electronic speed regulator (CR7513–E102).

21. MULTICOLOR PRINTING 318

    Color-register control (CR7505–W116)—Scanner circuits—Web-break relay—Speed relay.

22. PAPER-MACHINE CONTROL 331

    Multiple-generator system—Reference-voltage supply (CR7590–AD102)—Amplidyne pre-amplifier unit (CR7513–K102)—Timed acceleration—Tension control.

INDEX 345

PART I
RESISTANCE-WELDING CONTROL

CHAPTER 1

BASIC WELDING CONTROLS

In recent years electron-tube equipments have been used extensively in industrial plants. Of these equipments, two important groups are the controls for resistance-welding machines and the electronic speed controls of motors driving machine tools and industrial processes. Many plants use both of these types of tube-operated equipment.

In Part I this book describes many welding controls developed since 1943; it assumes that you may be familiar with the general background of resistance welding and electron tubes as outlined in earlier texts.*

1–1. Resistance Welding. Two pieces of metal may be welded or fused together by passing large current (1000 to 100,000 amperes) through these pieces while they are being forced together between the electrodes of the welding machine.†

A spot welder is shown in Fig. 1A. Figure 1B shows how the 220- or 440-volt a-c supply passes through a protective device, then through a contactor, before reaching this welding machine. In the machine, a welding transformer reduces the voltage at the electrode tips to 1 to 10 volts, and supplies the large welding current, while drawing perhaps 50 to 2000 amperes from the a-c supply. To make a weld, current needs to flow for only part of a second;‡ the contactor must close and open the circuit quickly,

---

* For quick reference to background material, many footnotes in this book will mention chapters in the following:

Chute, G. M., "Electronic Control of Resistance Welding," McGraw-Hill Book Company, Inc., New York, 1943.

Chute, G. M., "Electronics in Industry," McGraw-Hill Book Company, Inc., New York, 1946.

† When pieces of metal are being joined or welded together by a bright electric arc, which melts a rod into a pool of metal, that is called arc welding. No arc-welding equipment is described herein.

‡ To make a weld, the required heat $H = I^2RT$, or heat equals (current) $\times$ (current) $\times$ (resistance between the pieces welded) $\times$ (time while current flows). Since there must be resistance to current flow between the metal

and it does this hundreds of times each hour. While magnetic contactors control many such welders, ignitron contactors and other electron-tube equipments are used where better welds must be made in shorter time, with less contactor noise and maintenance.

Fig. 1A. A welding machine and its controls.

To make a single spot weld, the pieces of metal are placed in the space between the two electrodes, one of which can move. When the operator presses the button or the foot switch, the electrodes come together and squeeze the metal pieces. Welding cur-

---

pieces, where most of the weld heat is produced, we call this process *resistance welding*. This resistance depends on the metal that is being welded; steel has high resistance, so welding heat is easily produced; aluminum has low resistance, and the welding heat is harder to obtain. Further, this resistance between the metal pieces decreases when they are forced together by the electrodes with greater pressure.

rent then flows to heat the metal and make the weld. The metal is held under pressure for a moment until the weld hardens, then the electrodes separate so that the metal can be moved before the next weld is started. This complete welding operation may be controlled automatically by a sequence weld timer, as described in Chaps. 2 to 5.

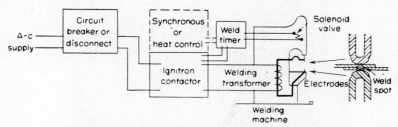

FIG. 1B. Welding machine and its electrical equipment.

1–2. Ignitron Contactor.* As is shown in Fig. 1B, an ignitron contactor may control or switch the alternating current supplied to the welding transformer. Such a tube contactor is shown in Fig. 1C. The cross section of one of the ignitron tubes is shown in Fig. 1D.

The ignitron is a high-current tube and has no internal heat from a filament (such as is used in a small radio tube). The ignitron uses a pool of liquid mercury as its cathode (like most mercury-arc rectifiers), so its enclosure contains mercury vapor. To fire the ignitron so that it passes large current through the space between its anode and cathode, a smaller current first must flow in its ignitor. A tiny arc then forms where the ignitor touches the mercury pool, and huge quantities of electrons† are driven up out of the surface of the mercury pool and are attracted to the large carbon anode whenever that anode is much more positive than the pool. Since its electrons flow in only one direction, cathode to anode, the ignitron is a rectifier. A single ignitron passes current during one-half wave of the a-c supply, so two ignitrons are needed, in place of a single switch contact, to close and open one side of the a-c supply circuit. As is shown in Fig. 1E,

---

* The ignitron tube and contactor are described in Chute, "Electronic Control . . .," Chap. 3, and in Chute, "Electronics in Industry," Chap. 9.

† An electric current may be considered as a flow of electrons, which are negatively charged particles that flow toward a more positive point in the circuit. In this book, all circuits are traced in the direction of electron flow.

6   ELECTRONIC MOTOR AND WELDER CONTROLS

Fig. 1C. An ignitron contactor.

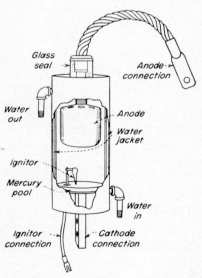

Fig. 1D. Ignitron tube cut away to show inside.

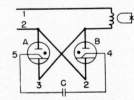

Fig. 1E. One contact C may fire both ignitrons.

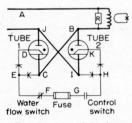

Fig. 1F. Circuit of ignitron contactor.

the two tubes are connected "back to back" so that the anode of one tube connects to the cathode of the other tube. A single control contact C may fire both tubes. With C closed and line 2 positive, electrons flow from line 1, through the load to point 3, from the mercury pool of tube A into its ignitor to 5, through contact C and into ignitor 4, to the mercury pool of tube B, to point 2 and line 2. This causes the arc between cathode 3 and ignitor 5, and "fires" tube A, which immediately passes electrons from cathode 3 to anode 2. Although this ignitor current fires tube A in normal fashion, these same electrons flow in reverse direction in tube B, or from ignitor to pool. Such reverse current definitely damages the ignitor and decreases tube life. To prevent this reverse current, copper-oxide (metallic) rectifiers are added, as shown in Fig. 1F. Each rectifier is a stack of copper disks, arranged so that they have very high resistance to electron flow in the direction indicated by the symbol arrow but low resistance to opposite electron flow.

Figure 1F shows the circuit used in a typical ignitron contactor. To fire ignitron tube 1, electrons first flow through path $ABCDEFGHIJ$. This current, flowing through ignitor D, causes tube 1 to pass electrons which flow $ABCJ$. During the following half cycle, ignitor electrons flow first through path $JIKHGFECBA$, which fires tube 2 so that load electrons then flow $JIBA$.

The control circuit $EFGH$ carries each ignitor's current in turn. Although this current must reach 25 to 40 amperes momentarily to fire the ignitron, it flows such a small portion of each cycle that a 3- or 6-ampere fuse serves during normal operation. If an ignitron fails to fire or becomes "hard starting," the ignitor current flows a larger portion of each cycle, and blows the fuse. In the same circuit, the flow-switch contact opens where there is not enough water flowing to cool the ignitrons.

Both ignitrons fire so long as the control switch is kept closed. For most tube contactors this control switch is a contact operated as part of a separate unit, such as some form of sequence weld timer.

1-3. Measuring the Weld Time. A simple kind of weld timer* is shown in Fig. 1G. Its time dial may be turned to select the length of time during which its relay is picked up or energized (usually measured in cycles of the a-c power supply); the relay

* This is classed as a NEMA 1A timer. Such standards of the National Electric Manufacturers Association are mentioned in Chap. 2.

8   ELECTRONIC MOTOR AND WELDER CONTROLS

contact is used to close the circuit that fires the tubes of the ignitron contactor to let welding current flow.

The circuit of this weld timer appears in Fig. 1H. At the top, transformer 1T furnishes a-c power; when the starting switch closes, this will let tube 2 pick up relay CR. Before the starting circuit is closed, no current can flow through either tube anode or through 2R or CR; therefore the CR contact is open. Meanwhile, in the grid circuit of tube 1, capacitor 1C has become charged by electrons flowing from point 9 through 6R, from cathode 6 to grid 11, through 1P and 4R to point 5. This voltage across 1C

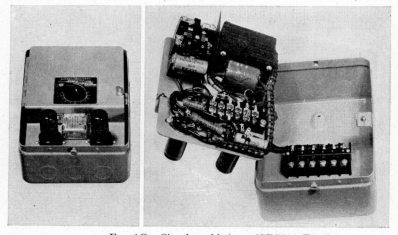

Fig. 1G. Simple weld timer (CR7503–F220).

keeps grid 11 more negative than point 5. Also, since there is no current flowing through 2R, there is no voltage or charge on 2C, and grid 15 of tube 1 is at the same potential as its cathode 9. When the starting switch closes, tube 2 has no grid bias;* electrons flow instantly from 9 through tube 2 and CR to 6, and relay CR closes its contact. However, although full anode voltage now appears across tube 1, the charge on 1C holds the grid 11 so negative that tube 1 passes no current. During this weld time, 1C gradually discharges through 1P and 4R. After the desired time delay set by 1P, electrons flow from 6 through tube 1, 7R and 2R to 9. This flow through 2R produces a voltage that charges

* A grid bias is a voltage, usually d.c., which keeps a tube from passing anode current; a "turn-on" signal may overcome this bias.

2C so that 14 is perhaps 70 volts more negative than 9; this voltage across 2C does not become less than 50 volts during the half cycle when tube 1 passes no current (for the time constant* of 2R and 2C is 0.05 sec., or 3 cycles). This voltage across 2C holds grid 15 so negative that tube 2 no longer passes anode current. Therefore, when tube 1 passes current, the resulting voltage across 2C prevents further flow through tube 2; CR drops out, opening the CR contact and stopping the welding current.

Notice that anode current flows in tube 1 only when 9 is more positive than 6; anode current flows through tube 2 only when 9 is more negative than 6. The voltage across 2R caused by tube 1 is stored in 2C and a half cycle later is used to control tube 2.

Additional types of weld timer will be discussed in the next chapter. Any of these timers will fire an ignitron contactor by closing a contact in the ignitor circuit.

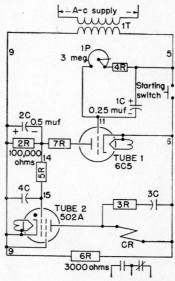

Fig. 1H. Circuit of a weld timer (CR7503–F220).

1–4. Thyratrons for Gradual Control of Ignitrons. When a magnetic contactor or an ignitron contactor closes the circuit to a welding machine, the operator may need to change taps on the welding transformer to obtain more or less weld heat as desired. However, when using an ignitron contactor it is easy to change the weld heat gradually merely by turning a small dial; the ignitrons must be fired by a pair of thyratron tubes in an added piece of equipment called a phase-shift heat control. This heat change may be made even while the ignitrons are passing current to the welder.

* The time constant (a term often used in electrical engineering) here refers to the length of time needed for capacitor 2C to discharge through resistor 2R, down to the point where the remaining voltage across 2C is about one-third of its starting voltage. This length of time is found by multiplying R (resistance in megohms) and C (capacitance in microfarads); sometimes it is called the RC time constant. In Fig. 1H, since 2R is 100,000 ohms or 0.1 megohm and 2C is 0.5 mu f (or mfd or μf), the RC time constant of 2R and 2C is 0.1 × 0.5 = 0.05 second. See Chute, "Electronics in Industry," Chap. 4.

10   ELECTRONIC MOTOR AND WELDER CONTROLS

In Fig. 1*I* we see thyratrons* 1 and 2 connected back to back in the ignitor firing circuit of an ignitron contactor. When the

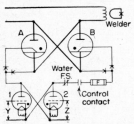

Fig. 1*I*. Thyratrons added to ignitron contactor.

control contact closes, ignitron *A* does not fire until thyratron 2 first passes anode current. So long as voltages *Y* and *Z* hold the grids sufficiently negative, no tubes will fire. If a-c grid voltage, applied at *Y* and *Z*, becomes positive early in the half cycles, tubes 1 and 2 each will fire its ignitron for entire half cycles of the power supply. However, if we can shift the phase or position of this a-c grid voltage, tubes 1 and 2 can be made to fire later (within each half cycle of a-c applied anode voltage) thereby decreasing the average amount of current that passes through ignitrons *A* and *B* and the welder. The welder current or heat is controlled in this way, by the complete phase-shift control next described.

1–5. Phase-shift Heat Control (CR7503–D175). In Fig. 1*J* this complete heat-control accessory is shown connected to an ignitron contactor. The two FG–95 thyratron tubes, 1 and 2, may fire the ignitrons in the same way as in Fig. 1*I*. However, these thyratrons are controlled by the a-c grid voltage supplied by transformer windings *S2T*, whose primary winding *P2T* is in the upper part of Fig. 1*J*. This phase-shifting circuit (5*R*, 4*R*, 4*C*, 3*R* and *P2T*) receives a.c. from a 230-volt winding of transformer 1*T*, having point 7 as its midtap.† The timing relay *TR* receives 115 volts as soon as power is applied; five minutes later, when tubes 1 and 2 have had enough filament heat, the *TR* contact closes. The starting contact now may fire thyratrons 1 and 2 and their ignitrons, to let welding current flow.

The voltage waves controlling the firing of thyratron 1 are shown in Fig. 1*K*. With the heat-control dial 5*R* turned so all

---

* The thyratron has a heated cathode; it is a vapor-filled tube, as shown by the dot inside the tube circle. Usually it is a negative-control tube, in that it may fire or permit anode current to flow while its control grid still is several volts negative. Still more negative grid potential prevents all current flow. When anode current starts, the grid cannot stop this current, which flows until after the anode voltage has been removed or reversed.

† Transformer 1*T* has more windings, not shown in Fig. 1*J*, with taps that permit the use of other supply voltages such as 200 volts or 480 volts, while holding 230 volts between points 6 and 8.

its resistance is in circuit, part (a) shows that the wave of S2T grid voltage remains negative until very late in the half cycle of tube-1 anode voltage. Tube 1 is fired at F; its anode current flows during a small portion of the half cycle, producing low welding heat.*

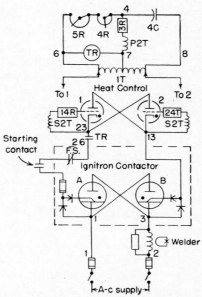

FIG. 1J. Welder heat control added to an ignitron contactor.

This condition is shown also by the vector diagram† in (b) of Fig. 1K. Across the voltage supplied by 1T (line 6–8, which is also the anode voltage of tube 1), the voltage drop across the large resistance (5R + 4R) is in phase with the current flowing in 4R; this current is 90 degrees ahead of the drop across capacitor 4C. The voltage applied to the primary winding P2T is seen to lead the 6–8 voltage; the secondary S2T voltage may be in phase with P2T or, by reversing the leads, S2T becomes directly opposite to P2T. As shown in (b), the S2T vector is lagging the 6–8 anode voltage by the angle G.

If 5R now is turned clockwise to increase the weld heat, (c) shows

* Note that the S2T voltage is large, so that its wave is nearly vertical at F; therefore this firing point is less affected by changes in line voltage or by differences between tubes.

† Chute, "Electronics in Industry," Chap. 13.

that the wave of $S2T$ voltage moves to an earlier position (toward the left) in the half cycle of anode voltage, firing tube 1 at J; tube-1 current flows for a larger part of the half cycle and becomes greater. As shown in (d), the amount of resistance ($5R + 4R$) is decreased; the current in $4R$ is greater and leads by a greater angle H. Since $P2T$ leads more, $S2T$ lags less (angle K).

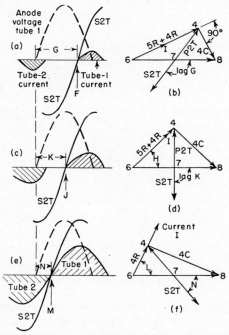

FIG. 1K. Wave shapes and vectors for phase-shifting circuit of Fig. 1J.

For the condition shown in (e) of Fig. 1K, heat control $5R$ is set for greatest heat; only the resistance of $4R$ remains, adjusted to give the lag N. Notice the full sine wave of current in tubes 1 and 2; the angle of lag N is the power-factor angle of the complete welder. The welding current is not increased further if tube 1 is fired earlier than M. In (f) the current in $4R$ and $4C$ has increased, leading by angle L. The $S2T$ vector is nearly in phase with the tube-1 anode voltage, lagging by the angle N.

When this phase-shift heat control is used with a welding transformer having taps, it is better to use a low transformer tap if possible, so that the heat dial always is set higher than 70 or 80 per cent. By using power during all of each cycle, the feeder voltage drop caused by that welder is decreased.

1-6. Load Ratings of Ignitron Tubes. In a welding control, the ignitron tubes easily may be overloaded when the welding machine is used for a heavier weld or a longer time than first intended. Figure 1L shows the rating curves for each of the four sizes of welder ignitrons. At 220 volts (left-hand chart) these tubes often may carry greater current than at 440 volts (right-hand chart). However, when operating continuously (at 100 per cent duty, shown at right-hand edge of each chart) the tubes have the same rating at both voltages. Size-D ignitrons are the largest now available for welding service; several pairs may be combined as described in Sec. 1-7.

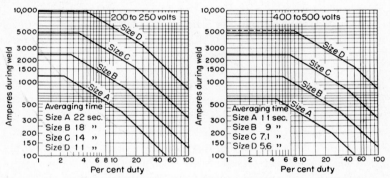

Fig. 1L. Rating curves of ignitron tubes (a pair) at 230 and 460 volts.

Figure 1L shows that a pair of size-B ignitrons may carry 130 amperes continuously, or 400 amperes at 30 per cent duty. In 60-cycle service, for example, this 400-ampere current may flow to make two welds per second, each weld being 9 cycles long. Here current flows 18 cycles out of each 60 cycles, or 30 per cent of the time. However, these same tubes must not be expected to carry this 400-ampere current for a weld 9 sec long, twice per minute, even though 18 sec/60 sec is 30 per cent. Notice in Fig. 1L that the averaging time of size-B tubes is 18 sec at 220 volts but only 9 sec at 440 volts. This means that the operation of size-B tubes, at 220 volts, must be figured within a total time not greater than 18 sec. Since a 9-sec weld is 50 per cent of 18 sec, these tubes are working at 50 per cent duty; here Fig. 1L shows 250 amperes as the highest load. Similarly, in a 440-volt circuit, the operation of these same tubes must be figured within 9 sec.

If a single weld is 9 sec long, the tubes are working during the whole averaging time, or at 100 per cent duty; their load should be only 130 amperes. This holds true even if the welder rests a minute or an hour between welds.

Then will size-C tubes handle this 400-ampere load for a 9-sec weld? At 220 volts the size-C averaging time is 14 sec. Since 9 sec/14 sec is about 65 per cent duty, Fig. 1L shows that nearly 500 amperes may be carried; size-C tubes are large enough at 220 volts. However, at 440 volts the size-C averaging time is only 7.1 sec. To make a 9-sec weld, the tubes are firing during the whole 7.1 sec so they are working at 100 per cent duty. Since Fig. 1L permits only 330 amperes load for size-C tubes at 100 per cent duty, the 400-ampere 9-sec weld will overload size-C tubes in a 440-volt circuit.

1–7. Ignitron Controls for Larger Loads. If a new welding machine will require, say, 6000 amperes from a 440-volt supply feeder, Fig. 1L shows this current to be greater than size-D tubes can carry even at very low duty. Nothing is gained by connecting together several pairs of size-D tubes and firing them all at the same time. Ignitrons do not divide the total current in the manner of high-vacuum tubes; whichever ignitron is first to fire will carry the whole load, sharing none of it with another tube in parallel.* For the 6000-ampere load above, it is best to supply two separate welding transformers,† each controlled by its pair of ignitrons carrying 3000 amperes (at less than 18 per cent duty, from Fig. 1L).

Let us consider another heavy welding load, such as 2000 amperes for 3.8 seconds at 440 volts. Here size-D tubes have 5.6 sec averaging time, and they would be working at 3.8 sec/5.6 sec or 68 per cent duty. The tubes are badly overloaded, for Fig. 1L shows that size-D tubes must not work above 35 per cent duty at 2000 amperes. But why not use two pairs of size-D tubes arranged so that each pair carries the 2000 amperes for half of the weld time, as described below? In this way each pair of tubes works 1.9 sec/5.6 sec or at 34 per cent duty; this is within the rating of size-D tubes.

---

* Ignitrons sometimes may divide large current loads if reactors are added in their anode circuits.

† The secondary windings of these two transformers should not be connected together; their separate sets of electrodes should not be close together where they press onto a single piece of work.

1-8. Flip-flop Control (CR7503-G119). A separate piece of equipment may be added to two ignitron contactors and a single heat control, so as to fire the tubes of each contactor in turn for about 2 sec each. In this way two pairs of size-D tubes may share a heavy load by dividing the weld time between them. This added control includes several relays, shown near the center in Fig. $1M$, which "flip-flop" in and out every 2 sec during the entire weld time. Relay $2CR$ connects the heat control to the first ignitor circuit so as to fire ignitrons A and B; 2 sec later, relay $3CR$ connects the heat control to the other ignitor circuit so as to fire ignitrons C and D, and $2CR$ drops out.* At the right in Fig. $1M$ the two pairs of ignitrons are connected in parallel, so each pair in turn passes current into the single large welding transformer.

The flip-flop control receives its a-c power through transformer $1T$, at the left in Fig. $1M$. Secondary windings $S1T$ heat the filaments of thyratron tubes X and Y; higher up, $S1T$ also furnishes 115 volts for picking up relays $1CR$ and $4CR$. The a-c power must be applied for about 30 seconds, to heat the tubes, before the flasher-button timer TD closes its contact, picking up relay $1CR$; this relay is sealed in by its contact 28-30.

In a lower circuit, tube Z rectifies the $P1T$ voltage which, filtered by reactor choke $1X$ and capacitor $1C$, becomes a supply of d-c voltage.† When the $1CR$ contact connects $1X$ to point 10, this d-c voltage makes point 10 about 200 volts more positive than point 3, connected to the cathodes of tubes X and Y. Through resistors $1R$ and $2R$, this d-c voltage appears between anode 12 and cathode 3 of tube X and also across tube Y; each thyratron will fire whenever its grid permits, as explained later. Notice that when tube X fires, electrons flow from cathode 3 to anode 12 and through $1R$ to 10; at this instant, if the $4CR$ contact is closed (as when relay $4CR$ is picked up by the starting contact in the welder or in the sequence timer), relay $2CR$ is energized also, closing the $2CR$ contacts to fire ignitrons A and B. When tube X is firing, its anode 12 is about 15 volts more positive than

---

* The contacts overlap; the $3CR$ contacts close before the $2CR$ contacts open. When these contacts operate during a continuous weld, current flows to the welder without interruption.

† For such rectifier and filter circuits see Chute, "Electronic Control . . .," Chap. 16, and Chute, "Electronics in Industry," Chaps. 2 and 10. See also footnote on page 24.

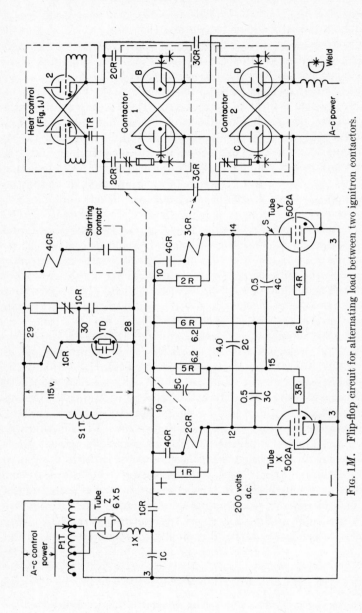

Fig. 1M. Flip-flop circuit for alternating load between two ignitron contactors.

its cathode 3; this 15-volt arc drop remains about the same, whether or not $4CR$ and $2CR$ happen to operate.

After $1CR$ has closed its contacts in Fig. $1M$, tube X fires first (thanks to tiny capacitor $5C$), then tube Y fires, turning off tube X; these tube circuits alternate every 2 seconds, as determined by the sizes of $3C$ and $6R$, $4C$ and $5R$. This timing action may be seen in Fig. $1N$, which shows the voltage wave shapes in various parts of the circuit. To hold this circuit still for a moment while we study it, let us open the anode circuit of tube Y, as if a switch were opened at S. In this condition, as shown at the left in Fig. $1N$, no electrons flow in $2R$; so point 14 is at the same positive potential as point 10. Electrons flow from cathode 3 to the control grid of tube X, then through $3R$ and $5R$ to point 10; since this grid is positive, tube X is firing (and $2CR$ is picked up if $4CR$ is energized to make a weld). Since anode 12 is only 15 volts above cathode 3, the large capacitor $2C$ has a large voltage charge that is $(+)$ at 14, $(-)$ at 12. Point 15 also is about 15 volts above 3,* so timing capacitor $4C$ is charged to the same voltage as $2C$. Meanwhile point 16 likewise is about 15 volts above tube-Y cathode 3, so both ends of $3C$ are at the same potential; $3C$ has no charge. The grid of tube Y is positive, so tube Y is ready to fire at once when its anode circuit is reclosed.

1–9. Thyratrons Alternating, on D.C. At J in Fig. $1N$ we let tube Y fire (by reclosing the anode circuit at S, its true condition); relay $3CR$ picks up. At once the potential at anode 14 drops to K (15 volts arc drop above 3), so terminal 14 of $2C$ likewise drops. Since a capacitor cannot lose its charge instantly (unless shorted), this charge on $2C$ forces the potential of its terminal 12 down to L; in this way anode 12 of tube X is driven negative long enough to stop the anode current, deionize the tube and permit its grid to regain control. At the same time the charge on $4C$ forces point 15 down to the negative potential at M; grid current stops and the tube-X grid remains negative for several seconds. This negative grid keeps tube X from firing, even though capacitor $2C$ quickly recharges (by electrons flowing up through tube Y into $2C$, and from terminal 12 through $1R$ or $2CR$ coil to 10), thereby letting anode 12 again become positive. While tube X is shut off, no voltage remains across $1R$; relay $2CR$ drops out. The new charge

---

* The ohm value of $3R$ is very small compared with that of $5R$, so the voltage across $3R$ is negligible; the arc drop holds the tube-X grid about 15 volts above the cathode.

on $2C$ is reversed; terminal 12 now is (+) and 14 is (−); $3C$ has charged to this same voltage.

While the $3CR$ contacts fire ignitrons C and D, there is a time delay of 2 seconds while $4C$ loses its charge,* letting point 15 rise

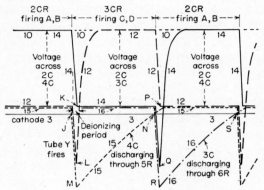

Fig. 1N. Wave shapes in timing circuit of Fig. 1M.

gradually from M to N. At N, point 15 has raised the tube-X grid close to cathode 3 and high enough to fire tube X and pick up $2CR$. At once anode 12 drops again, at P, so the charge on $2C$ now forces point 14 negative to Q, shutting off tube Y and dropping out $3CR$. This time the charge on $3C$ forces point 16 to the negative potential at R, so tube Y will not refire when its anode 14 quickly returns positive to the potential at 10. Capacitors $2C$ and $4C$ recharge. Again there is a 2-sec delay while $3C$ discharges through $6R$, letting point 16 rise from R to S. At S tube Y fires again, picking up $3CR$; its anode 14 drops and the $2C$ charge again turns off tube X. In this manner these two thyratrons X and Y will continue to fire in turn. At the end of the weld, when $4CR$ drops out preventing further operation of $2CR$ and $3CR$, tubes X and Y still fire at 2-sec intervals.

When taking turns every 2 sec, size-D ignitrons may be overloaded during a higher current short-time weld (such as 3500 amperes at 440 volts, flowing for 1½ sec, once every 8 sec). For

---

* As $4C$ changes its charge, electrons flow from its terminal 15 through $5R$ to 10. The potential at 15 rises at a rate set by the RC time constant ($5R \times 4C$, which is 3.1 sec). When point 15 has risen about 15 volts above cathode 3, tube-X grid current again flows through $3R$, preventing 15 from rising higher.

BASIC WELDING CONTROLS

such service, the G119 flip-flop control may be furnished with time-delay action of only ⅓ sec (or 20 cycles); relays 2CR and 3CR operate several times each second,* firing the ignitrons for shorter periods and letting them rest more often.

* To produce this faster action, 3C and 4C are changed to 0.05 mu f; 5R and 6R become 10 meg.

CHAPTER 2

G–E SEQUENCE WELD TIMERS

An ignitron contactor may pass current to cause a weld, but only while its ignitor firing circuit is closed. To fire the ignitrons at the right time, a sequence weld timer may be used. Such a timer controls also the solenoid valve that applies pressure to bring the welder electrodes onto the work. When the operator presses the starting or initiating switch, he starts the sequence weld timer, which makes the electrodes come together, pass welding current and then separate again.* Such timers are described in Chaps. 2, 3 and 4.

2-1. NEMA 3B and 5B Timers.† These types of sequence weld timer are used to fire ignitron contactors; each includes a timing circuit that determines how long the ignitrons shall fire to make a weld. The NEMA 3B timer is the most common type and is used to produce a simple spot weld; it includes four dials that separately control the squeeze time, weld time, hold time and off time, as shown in Fig. 2A. The NEMA 5B timer may produce a pulsation weld or a simple spot weld; it has six dials that control squeeze time, heat time, cool time, weld interval, hold time and off time, as shown in Fig. 2B.

2-2. Squeeze, Weld, Hold, Off. When the operator closes the starting switch,‡ the welder electrode tips come together quickly,

---

* Most weld timers or sequence controls are nonbeat. This means that a complete welding sequence or operation is carried through, even when the foot or initiating switch is closed and immediately opened again. A timer is sometimes arranged for use with a two-stage foot switch. Pressing such a switch part way lets the welder electrodes come together without passing any current. When pressed all the way, the switch causes complete welder operation.

† Such timer numbers have been adopted by the National Electric Manufacturers Association. Throughout the welding industry, each type number shows what timing circuits are included. See NEMA 1A in Sec. 1–3; NEMA 7B in Sec. 6–2; NEMA 9B in Sec. 6–8.

‡ In most welders, the starting switch closes the circuit to an electric solenoid valve, which lets air or hydraulic pressure be applied to a cylinder in the

but there must be a small time delay before welding current may pass safely. This time delay between the operator's signal and the start of welding current is called the *squeeze time*, for it is the time required for the electrodes to squeeze together properly. If current flows before the tips squeeze the work, there is too much heat at the tips, causing flash and burning of the work and tips.

At the end of the squeeze time (shown in Fig. 2A), the ignitron tubes fire and welding current flows. The total length of time, from the start of this current until the end of the current needed to make the weld, is called the *weld time*.

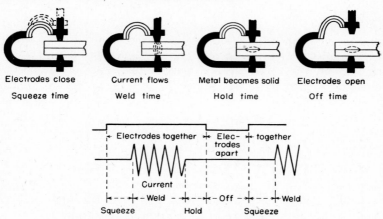

FIG. 2A. Sequence of a spot-welding machine.

While welding current flows, some of the work metal between the electrode tips melts for an instant. While in this fluid condition, the metal pieces are forced together by the pressure of the electrodes. When the welding current stops, the metal is still fluid for a short time. If electrode pressure is removed before the metal cools enough to regain its strength, the weld is weakened or cracked. The length of time between the stopping of welding current and the instant when the solenoid valve drops out, to remove electrode pressure from the work, is called the *hold time*. After the hold time, the electrodes separate, releasing the work. That weld is now finished.

Some welding machines are used to make a series of welds on the same piece of metal. The operator closes the switch for the

---

welder; the pressure in this cylinder moves the electrodes together so that the tips squeeze the work.

first weld, but desires the machine to make one spot weld after another so long as he keeps this switch closed. Between the end of one weld and the start of the next weld, the electrodes separate for part of a second. The length of time after the electrode solenoid valve opens until this valve recloses is called the *off time*. Many timers include a snap switch that can be set to give non-repeat operation and permit only single welds.

A NEMA 3B timer, having the four time delays explained above, is described in Secs. 2–8, 3–1, 4–1 and 4–11.

2–3. Pulsation Welding. Heat, Cool, Weld Interval. Most spot welders pass current only once while making one weld. Thicker pieces (such as ⅛ inch to several inches) are being welded by a method called pulsation welding, using a spot welder and a NEMA 5B timer. Instead of using one continuous flow of welding current, the single weld is made with a number of shorter impulses of current, separated by times when no current flows. In the example shown in Fig. 2B, each single weld may be made with three separate impulses of current, which flows here for 5 cycles during each impulse. Current stops for 4 cycles between impulses. After the third impulse, the weld is finished. The

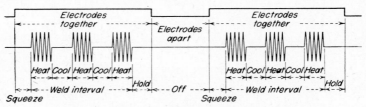

FIG. 2B. Operations during pulsation welds.

electrodes that had come together on the work before the first impulse still apply pressure during the hold time, and then separate for the off time before starting the next pulsation weld.*

Thick pieces may be welded by a long continuous flow of welding current, but this steady current also may produce such high temperature (at the place where the electrodes touch the steel pieces), that the electrodes soften or "mushroom," and press deep into the hot steel surface. During a pulsation weld the same amount of heat is produced at the electrodes but may be removed during the cool times so that the electrodes remain cooler and hold their shape.

* Chute, "Electronic Control . . .," Chap. 10.

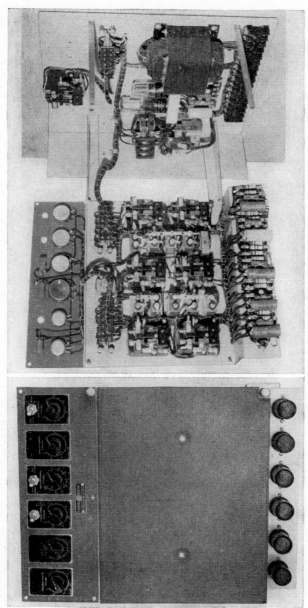

Fig. 2C. Sequence weld timer (NEMA 5B, G-E type CR7503-F180).

24 ELECTRONIC MOTOR AND WELDER CONTROLS

The length of each current impulse is called the *heat time;* between impulses, the length of the interval when no current flows is the *cool time.* The *weld interval* is the total time from the start of the first impulse until the end of the last impulse; therefore the weld-interval dial knob may be turned to get the desired number of current impulses for making a single weld.

2-4. Spot Welding with NEMA 5B Timer (CR7503–F180). The timer shown in Fig. 2C is intended for pulsation welding; its sequence of operation is described later as it times a pulsation weld. For now, in the elementary diagram of Fig. 2E, let us close switch 1S (at top center) so that this welding equipment produces merely a simple spot weld (a single impulse of welding current). Keep switch 3S open (upper right) to give a single nonrepeat operation; 3S keeps relay 6TD from operating.

As explained below in detail, after the squeeze time tube 1 picks up 1TD to fire tubes 3 and 4. Tube 3 fires tube 2 so that 2TD closes its circuit to fire the ignitron contactor and make a weld. After weld time, tube 2 drops out its relays, stopping the weld current and starting the hold time. Then tube 5 fires, letting the electrodes separate.

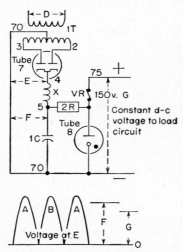

Fig. 2D. Circuit to supply steady d-c voltage.

Figure 2E shows that the anodes of tubes 1 to 6 and their relay coils 1TD to 6TD operate on a.c. (between points 1 and 10, ground potential) as supplied by transformer 1T at the right-hand side. However, the thyratron grid and timing circuits operate from the d-c supply between points 75 and 70. This d.c. is produced by the action of tubes 7 and 8, as described below* and in Fig. 2D.

---

* In Fig. 2D, transformer 1T supplies a-c voltage at anodes 2 and 3 of tube 7. During the half cycle A when anode 2 is more positive than center tap 70, electrons flow from 70 upward through the load (or through tube 8) to 75, through 2R and X, from cathode of tube 7 to anode 2. During the following half cycle B, when anode 3 is the more positive, electrons continue to flow from 70 through the load to 75, but from cathode to anode 3. Although

At the left-hand side of Fig. 2E, another secondary winding of transformer 1T furnishes an a-c voltage that is rectified by tube 7 and filtered by 1R and 1C. By means of voltage-regulator tube 8 (and its buffer resistor 2R) a steady d-c supply of 150 volts is held between the positive point 75 and negative point 70. Resistors 3R and 4R divide this 150-volt supply so that point 70 is always 90 volts more negative than 10 (ground), the cathode potential of tubes 1, 2, 4, 5 and 6. About 30 sec after the a-c supply is connected to 1T, all tubes are ready to operate; tube 8 is glowing. Before the starting switch is closed (at the right-hand side of Fig. 2E), no thyratron can pass anode current and no relay is picked up, for point 10 is not connected to transformer 1T.

Since the circuits near tube 1 are the same as those near tubes 4, 5 and 6, only one of these tube circuits needs to be explained.

The grid of tube 1 is connected, through a resistor and the closed 1CR contact, to 70; since point 70 is 90 volts more negative than the tube-1 cathode 10, tube 1 has a grid voltage of -90 volts. Through this closed 1CR contact, 150 volts d.c. is connected across resistors 11R and 12R; capacitor 11C is charged to this same voltage.

When the starting switch closes, it picks up relay 1CR; one

---

alternating voltage is supplied at D, a pulsating d-c voltage appears at E, consisting of half cycles, all above the zero line. To remove the unwanted ripple that remains of the a-c wave shape, a filter is used, here consisting of reactor choke X and capacitor 1C. The inductance of X tries to maintain a steady current through its winding, by storing energy during moments when current increases and then discharging this energy to help a decreasing current. (Sometimes a resistor, as 1R in Fig. 2E, replaces the reactor where less filtering is needed.) Similarly, capacitor 1C helps to smooth the voltage across the load by charging or storing energy during the high-voltage parts of the wave, and then discharging this energy during low-voltage periods. As a result, the rough voltage E becomes smooth at F.

However, the amount of this voltage F may change with every dip in supply voltage D. By adding tube 8 and resistor 2R we obtain a steady d-c voltage between points 75 and 70. Although voltage F may change, say, from 250 to 210 volts, load voltage G remains at 150 volts (since tube 8 is a 150-volt regulator tube; the gas inside it has a constant arc drop of about 150 volts). The "slack," or difference, appears across resistor 2R, varying between 100 and 60 volts in this example.

In Fig. 2D, the connection at VR is made by tube 8, by a jumper between two of its pins. If tube 8 is removed, this VR connection is opened, so that the load circuits do not receive excess voltage.

In some circuits, two or more voltage-regulator tubes are used in series with a single "buffer" resistor 2R (see tubes B and C in Fig. 15J).

See also Chute, "Electronics in Industry," Chap. 10.

26 ELECTRONIC MOTOR AND WELDER CONTROLS

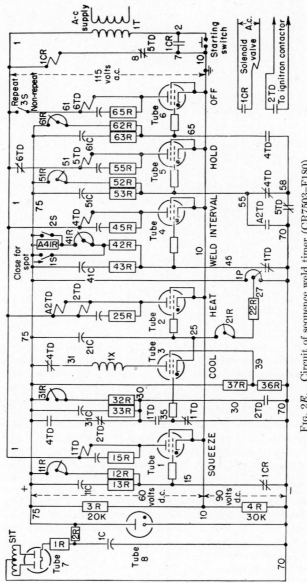

Fig. 2E. Circuit of sequence weld timer (CR7503–F180).

$1CR$ contact near the starting switch keeps $1CR$ energized. Another $1CR$ contact picks up the solenoid valve (lower right) to bring the electrodes onto the work. (All thyratrons now have voltage between the anode and the cathode, but they are kept from firing by their grids.) Another $1CR$ contact opens below tube 1, disconnecting point 15 from point 70. The potential of point 15 (grid of tube 1) now rises slowly, as capacitor $11C$ discharges through the combined resistance of $11R$, $12R$ and $13R$. Even if $11R$ is turned so that all its resistance is shorted, about 2 cycles of time pass before the voltage across $11C$ has decreased to approximately 60 volts, to let point 15 (grid) be near the same potential as 10 (cathode), so that tube 1 can fire. This time delay (before tube 1 fires) is the squeeze time; to increase this delay, more of the $11R$ resistance is put into circuit, so that $11C$ discharges more slowly.

2–5. Action of Tubes 2 and 3. During the squeeze time, before $1TD$ is picked up, the tube-3 control grid is held at -90 volts by $1TD$ contact (35 to 70). Since the other $1TD$ contact (35–30) is open, there is no charge on capacitor $31C$, so point 30 is at the positive potential of point 75. Meanwhile, notice that point 25 is the tube-3 cathode, yet it also controls the grid potential of tube 2. Tube 3 is not yet firing; if we overlook the shield-grid circuit 39 of tube 3, capacitor $21C$ will charge to the voltage between point 75 and slider 27 of $1P$.

However, shield grid 39 connects to a voltage divider ($36R$, $37R$) designed to keep point 39 about 15 volts below ground 10. When $21C$ terminal 25 tries to go more negative than point 39, electrons flow from 27 through $22R$ and $21R$, cathode 25 to shield grid* 39 and through $37R$ to 75. Since this holds point 25 near to point-39 potential, capacitor $21C$ is charged to the voltage between points 75 and 39, as shown at the left in Fig. 2F. This $21C$ voltage is also the voltage across tube 3, anode to cathode. Since the tube-2 grid is at the negative potential of point 39, tube 2 is not firing.†

When tube 1 fires after the squeeze time, relay $1TD$ switches the tube-3 grid from -90 to $+60$ volts, firing tube 3 at once.

---

* In most thyratrons, part of the shield-grid structure is between the cathode and the control grid; electrons may flow from cathode to shield grid even while the control grid is quite negative. This electron flow is very small so that it does not ionize the gas and cause undesired anode current.

† Heat-time-calibration adjuster $1P$ must be set so its slider 27 is more negative than point 10, or tube 2 may fire before the end of squeeze time.

28   ELECTRONIC MOTOR AND WELDER CONTROLS

The charge in capacitor 21C forces electrons to flow from cathode 25 through tube 3 and reactor 1X to point 75. As 21C discharges, its terminal 25 rises toward 75 (from A to B in Fig. 2F). This rising potential at grid 25 lets tube 2 fire when its anode next becomes positive (during positive half cycles from 1T); relay 2TD closes its contact, at lower right, to fire the ignitrons and start the flow of welding current. Another 2TD contact (30–70) connects the tube-3 grid to −90 volts. However, this vapor-filled

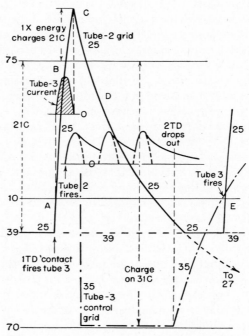

FIG. 2F.   Action of tubes 2 and 3 in Fig. 2E.

tube 3 continues to pass current, since its anode remains positive as 21C continues to discharge into 1X, storing energy in this reactor.

After point 25 has risen to the potential of point 75 (at B in Fig. 2F, where 21C has no charge), the tube-3 current continues while 1X releases its energy back into 21C. Now 1X is the supply that forces more electrons into terminal 75 of 21C; electrons return from 21C terminal 25, cathode to anode of tube 3, and back to 1X. Point 25 is driven more positive than 75, as shown at C. When

$1X$ has lost its stored energy, the tube-3 current stops, and the point-25 potential drops (as at D) as $21C$ slowly is charged by electrons flowing from point 27 through $22R$ and $21R$ into $21C$. When grid 25 becomes more negative than cathode 10, tube 2 drops out $2TD$, ending that impulse of welding current. The length of this current impulse or "heat time" is controlled by the setting of $21R$.

Meanwhile tube 4 has fired, since $1TD$ contact (45–70) opens after the squeeze time. With switch $1S$ closed for "Spot," $41C$ discharges very quickly through $42R$ and $43R$, so that tube 4 fires only a few cycles later than tube 1. One $4TD$ contact (31–75, above $1X$), opens quickly to prevent tube 3 from firing more than once; another $4TD$ contact (55–58) opens to start the hold time. After this delay set by $51R$, tube 5 fires; a $5TD$ contact drops out $1CR$ (at the right) to separate the electrodes. Since $3S$ is open, $6TD$ cannot operate; there is no further action until the starting switch is opened and reclosed for another weld operation.

2–6. Pulsation Weld Timing. If this sequence weld timer (Figs. 2C and 2E) is to produce a pulsation weld, switch $1S$ is open, and $41R$ is set for a weld interval long enough to permit the desired number of pulses of welding current. With this setting, the squeeze time and the first impulse of welding current will occur as described above; relay $1TD$ (contact 45–70) starts the weld interval, so that $41C$ slowly discharges through $41R$. During the first heat time, $2TD$ contact (30–70) is closed, so that $31C$ charges to 150 volts d.c. between points 75 and 70. When tube 2 drops out $2TD$ to stop the welding current after the heat time, the contact 30–70 opens, letting $31C$ discharge through $31R$, $32R$ and $33R$. After the cool time set by $31R$, points 35 and 30 have risen close to the potential of point 10 (as shown at E in Fig. 2F); if $4TD$ has not yet operated its contacts, tube 3 fires again, discharging $21C$ and firing tube 2 to cause another impulse of welding current (like that described in Sec. 2–5). The $2TD$ contact (30–70) again charges $31C$ and turns off tube 3. After this second heat time, $2TD$ again drops out, starting a second cool time. More welding-current impulses will occur, with cool times between them, until the end of the weld interval. Then, when tube 4 fires and picks up $4TD$, its contact 31–75 prevents further firing of tube 3. Another $4TD$ contact opens (55–58) but cannot start the hold time until the welding current stops, letting $A2TD$ open its contact 55–70.

2-7. Repeat Welding. Off Time. If switch $3S$ is closed (upper right in Fig. $2E$), the welding machine will make a complete weld, separate the electrodes during the off time, then start another complete welding operation, repeating this sequence of welds so long as the starting switch remains closed.

With $3S$ closed, relay $6TD$ is picked up when the starting switch is closed. When tube 4 fires, to end the pulses of welding current, a $4TD$ contact connects tube-6 grid 65 to the negative potential of point 58; $6TD$ drops out, closing its contact 1–51 above tube 5. Notice that $4TD$ and $2TD$ contacts (above tube 1) now let $31C$ discharge quickly through $33R$ to be ready for the next welding sequence even if $31R$ is set for a long cool time.

After the hold time, tube 5 picks up $5TD$ whose contact drops out $1CR$ to let the electrodes separate. Another $5TD$ contact opens (70–58) to start the off time; during this time $61C$ is discharging at a rate set by $61R$. If the starting switch still is closed at the end of off time, tube 6 picks up $6TD$ which drops out $5TD$; the $5TD$ contact 7–8 picks up $1CR$ to start another welding operation. Since $1CR$ seals itself in (7–2), the starting switch has no control action except during the off time; when once started, the welder performs a complete operation even if the starting switch is released immediately.

2-8. Sequence Timer Combined with Contactor (CR7503-L102A8). Many recent spot-welder controls include a $3B$ or $5B$ timer in the top of the case that holds the ignitron contactor; the controls described in Chap. 3 are arranged in this way. An example of such a combination is next described. As shown in Fig. $2G$, the lower portion is like the ignitron contactor of Fig. $1C$, explained in Sec. 1–2; the upper portion is a $3B$ timer that controls a welding machine through its squeeze, weld, hold and off times. While this timer has four time-adjusting dials on the outside of the cabinet, it uses only two small thyratrons in controlling these four time delays.

Between the two large ignitron tubes is a plug-in relay CR that is operated by the sequence timer above. When CR closes its contact, this fires the ignitrons so that current flows to make a weld. A thermal flow switch is included so that CR cannot pick up unless cooling water is flowing through the ignitrons.

Figure $2H$ shows the elementary diagram of the ignitron contactor combined with its sequence timer. The action of transformer $1T$ and tubes 7 and 8 has been described in Sec. 2–4 and in the footnote on page 24. They provide d-c voltage so that point 75 remains 60 volts above ground 7, while point 70 remains 90 volts below 7.

Fig. 2G. Combination sequence timer with ignitron contactor (CR7503-L102A8).

After voltage has been applied to the sequence timer for 30 seconds to warm the tubes, the equipment is ready to operate. Tube 8 on the back panel has a colored glow. Before the starting switch closes, thyratrons 1 and 2 pass no current; no relays are energized.

When the starting switch closes, tube 2 fires at once, picking up relays $2TD$, $A2TD$ and $1CR$, bringing together the welder

32   ELECTRONIC MOTOR AND WELDER CONTROLS

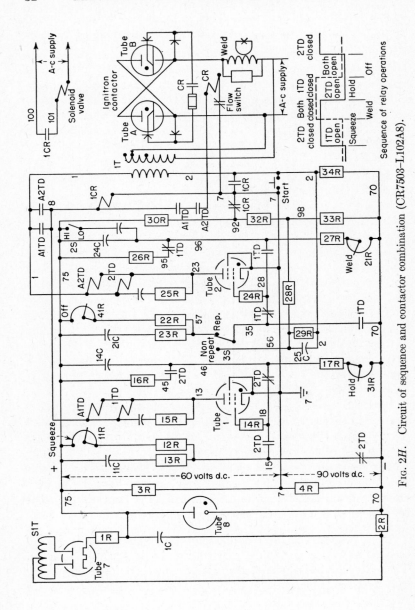

Fig. 2H. Circuit of sequence and contactor combination (CR7503-L102A8).

electrodes. After the squeeze time, tube 1 picks up relays $1TD$ and $A1TD$ so that relay CR fires the ignitron tubes. At the end of weld time, tube 2 stops firing, so relay $A2TD$ drops out CR, stopping the welding current. After the hold time, tube 1 also stops firing, so relay $A1TD$ drops out $1CR$ and the welder electrodes separate. If switch $3S$ is set at "Nonrepeat," tube 2 cannot fire again to start another weld operation until after the start switch has been released and reclosed. With $3S$ at "Repeat," tube 2 fires again at the end of the off time, bringing the electrodes together for another weld operation. This action is detailed below.

Before the starting switch is closed, relay $1CR$ is not picked up, so the a-c supply voltage ($1T$) is not connected to point 7. There is no voltage to force anode current through tubes 1 and 2, so relays $1TD$, $A1TD$, $2TD$ and $A2TD$ all are dropped out; their normally-closed contacts (⌗) are now closed. The control grid of tube 1 is connected through $14R$ and $2TD$ contact to point 46 which is now at the negative potential of point 70 (since capacitor $14C$ is charged to the entire 150 volts between 75 and 70). To the left of tube 1, a $2TD$ contact connects point 15 to 70; capacitor $11C$ is charged to the 150 volts 75-to-70. To the right of tube 2, a $1TD$ contact 95–96 has connected $26R$ in series with $27R$ and $21R$, to become a voltage divider between points 75 and 70. Since $26R$ has few ohms compared with $27R$, the voltage across $26R$ and $24C$ is small, so that point 96 is at least 35 volts more positive than point 7.

With switch $3S$ set for "Repeat," the control grid of tube 2 is connected through $24R$ and $1TD$ contact to point 35, and through $3S$, $22R$ and $41R$ to positive point 75. Tube 2 is ready to fire as soon as it receives a-c anode voltage.

2–9. Control Sequence, Repeat Weld. When the starting switch is closed, connecting points 2 and 7 (at the right in Fig. $2H$), tube 2 fires instantly, picking up relays $2TD$ and $A2TD$. The $A2TD$ contact 1–8 closes, picking up relay $1CR$ which seals itself in by its $1CR$ contact 2–7, so that the starting switch may be released. Another $1CR$ contact 100–101 closes, applying voltage to the solenoid valve, which brings the welder electrodes together. Another $A2TD$ contact closes, but cannot yet pick up CR. Above tube 1, a $2TD$ contact 45–46 connects $16R$ in series with $17R$ and $31R$ so that $14C$ discharges to a small voltage, and point 46 becomes at least 35 volts above point 7. To start the squeeze time, contacts of $2TD$ (shown under tube 1) now connect tube-1 grid to point 15; another $2TD$ contact disconnects 15 from 70. As capacitor $11C$ now discharges through $13R$, $12R$ and $11R$, the potential of 15 rises upward from 70; after the squeeze time, as selected by the setting of $11R$, point 15 (tube-1 grid) has come close to point 7 (tube-1 cathode) so that tube 1 fires, picking up relays $1TD$ and $A1TD$, to start the weld time.

One $A1TD$ contact completes the a-c circuit through $A2TD$ contact, to pick up relay CR of the ignitron contactor. This CR contact closes the circuit that fires ignitron tubes A and B, to produce heat at the weld. Below tube 2, contacts of $1TD$ now disconnect grid 28 from point 35 and connect it instead to point 96, which is positive so that tube 2 continues to fire. (Meanwhile another $1TD$ contact connects 35 to 70, charging $21C$ to be ready for the off time.) To start the weld time, the $1TD$ contact 95–96 opens, so that capacitor $24C$ now charges to greater voltage, driving 96 more negative because of electrons flowing from 70 through $21R$ and $27R$. The potential of point 96 drops at a rate set by the weld-time setting of $21R$. By decreasing resistance in $21R$, point 96 drops faster and the weld time is made shorter.*

At the end of this weld time, point 96 has become more negative than cathode 7, so that tube 2 cannot fire during the following cycles of its a-c anode voltage; relays $2TD$ and $A2TD$ drop out. The $A2TD$ contact opens, so relay CR opens the firing circuit of ignitrons A and B, and welding current stops. Below tube 1, contacts of $2TD$ now disconnect grid 18 from point 15 and connect it instead to point 46, which is positive, so that tube 1 continues to fire. (Meanwhile another $2TD$ contact connects 15 to 70, charging $11C$ to be ready for the next squeeze time.) To start the hold time, the $2TD$ contact 45–46 opens, so that capacitor $14C$ now charges to greater voltage, driving 46 more negative, because of electrons flowing from 70 through $31R$ and $17R$. The potential of point 46 drops at a rate fixed by the hold-time setting of $31R$.

At the end of this hold time, point 46 has become more negative than cathode 7, so that tube 1 cannot fire during the following cycles of its a-c anode voltage; relays $1TD$ and $A1TD$ drop out.

* If the high-low switch $2S$ is closed (for 90 cycles maximum weld time), a capacitor is added in parallel with $24C$. Since these capacitors together have three times the capacity (in microfarads) of $24C$ alone, point 96 drops one-third as fast; for any position of $21R$, the weld time is three times as long as when $2S$ is open.

The $A1TD$ contact 1–8 opens the a-c circuit to $1CR$; the $1CR$ contacts 100–101 remove voltage from the solenoid valve so that the welder electrodes separate. Below tube 2, contacts of $1TD$ now reconnect grid 28 to point 35 and (through switch $3S$, set for repeat) to point 57 which is negative because of the charge on $21C$. (Meanwhile a $1TD$ contact connects 95 to 96, partly discharging $24C$ and raising point 96 to be ready for the next weld time.) To start the off time, the $1TD$ contact 35–70 opens; as capacitor $21C$ now discharges through $23R$, $22R$ and $41R$, the potential of 35 and 57 rises upward from 70; after the off time as selected by the setting of $41R$, tube-2 grid 35 has come close to tube-2 cathode 7. If the starting switch still is closed, tube 2 fires, again picking up $2TD$ and $A2TD$, bringing the welder electrodes back together to give a repeat weld operation.

2–10. Nonrepeat Weld. If switch $3S$ is set for "Nonrepeat" (near the center of Fig. $2H$), $3S$ connects point 35 to 56, which is one terminal of capacitor $25C$. This $25C$ circuit provides a way to keep tube 2 from firing a second time, even though the starting switch may be held closed after the first weld is completed. During the weld and hold times when tube 1 is firing and the $1TD$ contact connects 35 to 70, notice that terminal 56 of capacitor $25C$ is at the negative potential of point 70, while the other terminal 2 of $25C$ is at the ground potential of point 7. At the end of the hold time, just as relay $1TD$ drops out, capacitor $25C$ has a 90-volt charge. While the starting switch is held closed, point 56 remains more negative than cathode 7. Because of this charge on $25C$, the tube-2 grid (connected through $24R$, $1TD$ contact, and $3S$ to point 56) is kept negative during the short time after $1TD$ contact 70–35 has opened but before relay $1CR$ drops out to close its contact 92–7 at the right. If $25C$ were not used, the potential at point 98 (on voltage divider $30R$, $32R$, $33R$) would refire tube 2, just before $1CR$ connects 92 to 7. However, after this 92–7 contact has closed, point 98 becomes about 20 volts more negative than 7.

When the operator releases the starting switch, point 2 is disconnected from ground 7 and drops to the negative potential of 70. Capacitor $25C$ now receives a 65-volt charge (from the voltage between points 98 and 70) but 56 is now the more positive terminal. When the starting switch is closed again, point 2 rises instantly to ground potential; the charge on $25C$ forces point 56 and tube-2 grid 28 far above cathode 7, to fire tube 2.

2–11. Two Starting Switches Select Two Weld Times.

As shown in Fig. 2*I*, an accessory control (CR7503–DY2) may be added to the timer of Fig. 2*H*, so that a single welding machine may be used for two different weld operations. This accessory includes a relay *DCR* and a second weld-time adjuster *A21R*, like 21*R* in the sequence weld timer.

If starting switch 1 is closed, the added relay *DCR* is not picked up; the sequence weld timer operates just as described previously,

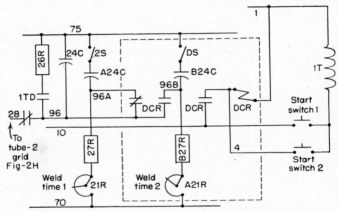

Fig. 2*I*. Accessory control for two weld times.

so that the weld time is set by 21*R*. During this weld time, the tube-2 grid is connected through 1*TD* contact to 96, through *DCR* contact to 96*A*; capacitor 24*C* charges through 21*R* and 27*R*. However, when the operator closes starting switch 2 instead, this picks up relay *DCR* so that the tube-2 grid is connected through 1*TD* and *DCR* contacts to 96*B*; capacitor 24*C* charges through *A21R* and *B27R*, so that *A21R* sets the length of weld time 2. Separate range switches 2*S* and *DS* are used so that weld time 1 may be much longer or shorter than weld time 2. For example, if weld time 1 is 10 cycles but weld time 2 should be 45 cycles long, switch 2*S* is open (for the range 3 to 30 cycles) but *DS* is closed (for range 30 to 90 cycles) so that weld time 2 is controlled by both 24*C* and *B*24*C* charging through *A21R* and *B27R*.

By the use of similar accessory controls most designs of welding timer may be arranged to provide two or more weld times, per cent heats or other dual adjustments.

CHAPTER 3

MORE SEQUENCE TIMERS

While the sequence timers in Chap. 2 include relays operated by tubes in the timing circuits, the sequence timers in Chaps. 3 and 4 include fewer relays; more thyratron tubes here take the place of relay contacts, so that the timer has fewer moving parts.

3-1. Weltronic 3B Timer (Model 75SE2). As shown in Fig. 3A, and the diagram, Fig. 3B, this timer uses 10 small thyratrons, 2 larger tubes 11 and 12, and a single relay (for the solenoid valve) to provide squeeze, weld, hold and off times.

Before starting a weld operation, only tubes 3, 5 and 7 are passing anode current. When the starting switch closes (at upper left), tubes 1 and 2 pick up relay $SVCR$ to bring the electrodes together; tube-3 current stops. After squeeze time, tube 4 fires, turning off tube 5 to start the weld time. Tubes 9 and 10 fire, thereby turning on the larger thyratrons 11 and 12; these fire the ignitrons to let welding current flow. After the weld time, tube 6 fires, turning off tubes 7, 9, 10, 11 and 12 to stop the ignitrons and the welding current. After the hold time, tube 8 fires, turning off tubes 1 and 2, dropping out $SVCR$ to separate the electrodes. Tubes 3, 5 and 7 turn on; tubes 4, 6 and 8 turn off. If the starting switch is held closed, and $2S$ is set for repeat welding, tubes 1 and 2 will fire again after the off time, starting another weld sequence. Let us now study this in detail.

Control power enters through transformer $1T$ (at upper left) to heat tube filaments and to supply 110 volts continuously to the small transformers $4T$, $5T$, $6T$ and $9T$; at all times these transformers supply 170 volts to the anodes of tubes 3, 4, 5, 6 and 7, which may fire whenever their grids permit. Before the starting switch is closed, notice that point 2 (cathode of most tubes) is not connected to terminal 1 of the 110-volt winding of $1T$; there is no voltage to force current through tubes 1 and 2, to pick up $SVCR$. Point 13 (top of tube 1) is at a potential near to point 4; therefore one winding of transformer $2T$ (between points 13 and 49 near tube 2) is receiving voltage (from transformer $1T$ terminal 1

Fig. 3A. Sequence weld timer, Model 75SE2. (*Weltronic Company.*)

through $5R$ to $2T$, then through $SVCR$ coil to 4). Thus energized, the other $2T$ windings produce voltages that drive tube-2 grid negative, but drive tube-3 control grid positive; tube 3 is passing current. Transformer $S4T$ forces electrons to flow from terminal 25 through $9R$ and $1P$ to 2, then cathode to anode of tube 3. Since only 15 volts now appears across tube 3, the rest of the $S4T$ voltage charges capacitor $2C$ so that its terminal 25 is more negative than 2. This voltage across $2C$ also is a grid voltage that prevents tubes 4 and 9 from firing. With no current through tube 4 and $15R$, there is no voltage across $3C$, so tube 5 has zero grid voltage and is passing current; this charges $4C$ to drive negative the grid 27 of tube 6. With no current through tube 6 and $17R$, there is no charge on $5C$, so tube 7 is passing current (and control grid 29 of tube 9 is at cathode-2 potential); the charge on $6C$ keeps tube-8 grid negative. With no current through tube 8, $4P$ or $23R$, there is no voltage across $7C$ so point 23 (at far right and connected to grids of tubes 1 and 3) is at cathode-2 potential. Both grids of tube 3 permit it to fire. Tube 1 is ready to fire as soon as voltage is applied across it by the starting switch.

Near the bottom of Fig. $3B$ notice that the lower half of transformers $3T$ and $8T$ receives voltage continuously from bottom line 1 and center tap 4 (which are the 110-volt output terminals of $1T$). As a result, so long as tube 9 is not firing, the secondary voltage of $3T$ prevents tube 10 from firing; similarly, the $8T$ secondary voltages are out of phase with the anode voltage of tubes 11 and 12 so that these tubes cannot fire the ignitrons.*

3-2. Weltronic Timer Operation. When the starting switch closes, it connects tube-1 cathode 2 to transformer-$1T$ terminal 1, so that electrons flow from 1 and 2 up through tube 1 to anode 13 and through $SVCR$ to $1T$ terminal 4. Since point 13 now is brought down to 15 volts above cathode 2, terminal 13 of transformer $2T$ is at potential lower than its center tap, which connects through $4R$ to $1T$ terminal 4. Therefore the firing of tube 1 reverses the phase of the voltage applied to $2T$; the $2T$ secondary voltages reverse, to turn on tube 2 and to drive negative the tube-3 control grid.† Tubes 1 and 2 (connected back to back) pick up

---

* The voltages across $9C$ and $10C$, charged by grid rectification, drive the grids negative before the anodes become positive, thereby preventing unwanted firing of tubes 11 and 12.

† Since the $SVCR$ coil is an inductive type of load, the tube-1 current continues to flow after the line voltage has reversed (as described later in Sec. 4–4);

relay $SVCR$. One $SVCR$ contact seals around the starting switch, which now may be released without stopping the weld operation; the other $SVCR$ contacts pick up the solenoid valve to bring together the electrodes onto the work.

Since the tube-3 current stops when the starting switch closes, capacitor $2C$ receives no further charge through tube 3; $2C$ now discharges at a rate set by squeeze-time adjuster $1P$. There is a squeeze-range switch $1S$ that shorts $10R$ for shorter timing; for longer squeeze time, electrons flow from $2C$ at 25 downward through $9R$, $1P$ and $10R$ to $2C$. As this $2C$ voltage nears zero, point 25 rises close to cathode 2 and fires tubes 4 and 9.

When tube 9 fires (lower part of Fig. $3B$) electrons flow from points 1 and 2 up through tube 9 to point 40, through half of $3T$ to point 4. Since anode 40 now is only 15 volts above cathode 2, the $3T$ primary is energized by the voltage 40 to 4 (that has phase opposite to 1–4); this reverses the voltage applied to the grid of tube 10, which fires during the half cycle following tube 9. If the weld–no-weld switch $3S$ is closed, connecting 40 to transformer $8T$, the firing of tubes 9 and 10 similarly reverses the voltage that energizes $8T$. Voltage reverses at each of the $S8T$ windings, becoming in phase with the anode voltages of tubes 11 and 12; these larger thyratrons instantly fire the ignitron tubes to let welding current flow.*

Meanwhile, when tube 4 fires (at the end of the squeeze time), electrons flow up through tube 4 and down in $15R$; this charges $3C$ so that the grid of tube 5 becomes negative. The tube-5 current stops, so that capacitor $4C$ may discharge through $2P$ and $16R$, at a rate set by weld-time adjuster $2P$. After this weld time, tube 6 fires, charging $5C$ so that grid 29 of tube 7 is driven negative, to start the hold time. Notice that point 29 is connected also to the control grid of tube 9. Although tube 9 was turned on by its shield grid at the start of weld time, it is turned off by its control grid at the end of weld time. As the tube-9 current stops, point 40 no longer is held at low potential, so transformers $3T$ and $8T$ again furnish secondary voltages that prevent tubes 10, 11 and 12 from firing. The ignitrons stop the flow of welding current.

---

the $2T$ voltages remain reversed long enough to fire tube 2 in the following half cycles. Similar action permits tube 9 and $3T$ to fire tube 10.

* This connection of thyratrons to ignitrons has been described in Sec. 1–5.

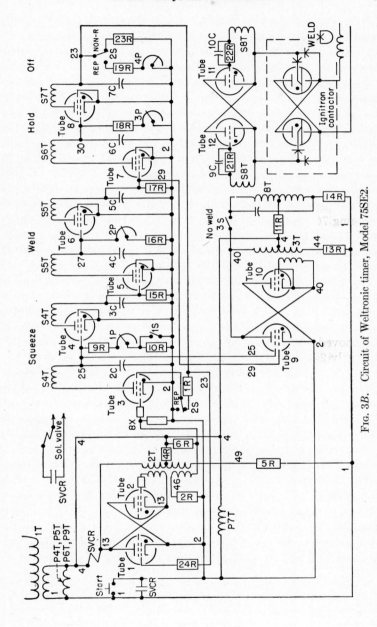

Fig. 3B. Circuit of Weltronic timer, Model 75SE2.

3-3. Single or Repeat Welds. When the tube-7 current stops, capacitor $6C$ discharges at a rate set by hold-time adjuster $3P$. After the hold time, tube 8 fires, its anode transformer $7T$ energized by the starting switch. If switch $2S$ is set for a single weld ("Nonrepeat"), electrons flow from $S7T$ through $23R$ and tube 8, charging $7C$ so that its terminal 23 becomes negative. This negative 23 potential now appears at the grid of tube 1 and, through $2S$, at the shield grid of tube 3. Tubes 1 and 2 drop out relay $SVCR$, letting the electrodes separate. If the starting switch has been released, the opening of $SVCR$ contact removes voltage from transformer $7T$, stopping the tube-8 current.* Quickly $7C$ discharges through the low resistance of $23R$; the negative potential is removed from the tube-3 shield grid, so tube 3 fires. Almost instantly $2C$ charges, turning off tube 4. Similarly, $3C$ discharges quickly, firing tube 5 to turn off tube 6; $5C$ discharges quickly, firing tube 7. The circuit is ready when the starting switch again closes to begin the next weld operation.

For repeat weld operation, put $2S$ in the repeat position. This connects the tube-3 shield grid to cathode 2. When tube 8 fires after hold time, electrons flow from $S7T$ through $19R$, $4P$ and tube 8, charging $7C$. Its terminal 23 becomes negative, turning off tube 1; relay $SVCR$ drops out and transformer $2T$ turns off tube 2 but makes positive the tube-3 control grid. As described above, tube 3 fires turning off tube 4 (and driving negative the shield grid of tube 9); tube 5 fires turning off tube 6; tube 7 fires turning off tube 8. Capacitor $7C$ now discharges at a rate set by off-time adjuster $4P$. After this off time, if the starting switch still is closed, the rising potential of point 23 lets tube 1 fire, to pick up relay $SVCR$ and repeat the weld operation.

3-4. Taylor Winfield Pulsation Sequence Timer (N3 Combination). This design of $5B$ sequence timer is mounted together with an ignitron contactor, all in one enclosure, thereby becoming a NEMA N3 combination. As pictured in Fig. $3C$, this timer uses 11 tubes and two relays; its diagram appears in Fig. $3D$. Alternating-current control power is supplied through the transformer at upper left, and is rectified by tube 9 to produce 275 volts d.c.

---

* If the starting switch still is closed, tubes 1 and 2 drop out $SVCR$ and let $2T$ make the control grid of tube 3 become positive. However, the charge on $7C$ still holds negative the tube-3 shield grid. Tubes 3, 5 and 7 cannot fire again and a new weld operation cannot begin until the starting switch has been released.

Fig. 3C. NEMA N3 combination timer. (*The Taylor Winfield Corporation.*)

between top point 1 and grounded point 2. At the lower left a metallic half-wave rectifier and capacitor $2C$ apply a d-c voltage across tube 11 and $40R$, so that bottom point 3 remains always 75 volts more negative than ground 2.*

Either a single-stage or a two-stage foot switch may be used.† To study first the most simple operation, let us use a single-stage switch $FS1$; it must have two contacts. An open $FS1$ contact is shown above tube 11 in Fig. 3D; a normally-closed $FS1$ contact is connected at B. Switch contacts are not used at A or C, but a jumper must be placed at A (as shown dotted). Close switch $S1$ (at lower left) for a single nonrepeat weld; set the impulse-spot switch $S2$ (at lower right) to the S position as shown, for a simple spot weld.

Before closing $FS1$, only tubes 9 and 11 are passing current. Neither relay $1CR$ nor $2CR$ is picked up.

Briefly, $FS1$ now will fire tube 1, to pick up $1CR$ and fire tubes 3 and 5. After squeeze time, tube 4 fires and, in series with tubes 3 and 5, picks up $2CR$ to fire the ignitron contactor. After the heat time tube 6 fires, turning off tube 5 and dropping out $2CR$. Immediately tube 7 fires, starting the timing action of the tube-8 circuit. After the hold time, tube 8 fires tube 2, turning off tube 1 and dropping out $1CR$.

3–5. Squeeze, Heat and Hold Times. Before closing $FS1$, note that capacitors $3C$ and $4C$ (near $FS1$) are charged through $42R$ and $43R$ to the d-c voltage between ground point 2 and positive point 49 on resistor $37R$. Also, through a n-c (normally-closed) contact of $1CR$ (1–70, at top center), resistor $15R$ is connected across the 275 volts d.c. between points 1 and 2, so that slider 61 on $15R$ has a potential of about $+200$ volts. Capacitor $11C$ is charged by electrons flowing from 2 through $1CR$ contact to 34 (at center of Fig. 3D), cathode to control grid of tube 4 and through $53R$, so that 62 is more negative than 61. At this time there is no anode voltage on tubes 3 to 8, since the normally-open contact of $1CR$ (at top) separates 60 from 1. Voltage-regulator tube 10 is not glowing. At top center, $25C$ is charged (through $9R$ and $10R$) to the voltage between points 1 and 2. With switch $S1$ closed for nonrepeat, the control grid of tube 1 (at point 19) is kept at the -75-volt potential of point 3.

When the operator closes $FS1$, this lets capacitor $3C$ discharge

---

* See footnote, p. 24.
† See footnote, Sec. 2–1.

through $P1T$; this transformer produces an impulse peak of 100 volts at $S1T$ that fires tube 1, picking up relay $1CR$; this operates the solenoid valve to bring the welder electrodes onto the work. Another $1CR$ contact closes (at the top, 1–60) discharging $25C$ through $P4T$, so that $S4T$ (right center) fires tube 5. Electrons flow from 2 through $10R$, cathode to anode of tube 5, $20R$ and $2CR$; this current is not large enough to pick up $2CR$. Another $1CR$ contact opens at 2–34, but closes at A; this lets $4C$ discharge through $44R$, $P3T$ and the dotted jumper; $S3T$ produces a 100-volt peak to fire tube 3, thereby holding point 34 about 15 volts above 2.

When the $1CR$ contact disconnects 70 from 1, terminal 61 of $11C$ drops close to 2. However, because of the previous charge on $11C$, its terminal 62 is driven far more negative than 2, keeping tube 4 from firing during the squeeze time.

When $1CR$ connects 1 to 60, electrons flow from 2 through tube 10 and $2R$ to 60 and 1; tube 10 now glows and point 50 rises to $+150$ volts. This positive potential is applied through $1P$ and $16R$ to terminal 62 of $11C$. Therefore $11C$ begins to lose its negative charge, by electrons flowing from 2 through $15R$ to slider 61 and from $11C$ terminal 62 through $16R$ and $1P$ to point 50. After this squeeze time, as set by $1P$, point 62 has risen close enough to point 2 so as to fire tube 4; point 27 (tube-4 anode and tube-5 cathode) drops to about 30 volts above 2.

An increased flow of electrons now passes from 2 up through tubes 3, 4 and 5, $20R$ and $2CR$ coil, and picks up relay $2CR$. One $2CR$ contact fires the ignitron contactor to let welding current flow. Another $2CR$ contact closes (59–66, at lower right), shorting $27C$ through $36R$ and driving negative the control grid 65 of tube 5; this cannot turn off tube 5 because the anode of this vapor-filled tube still is positive. The $2CR$ contact 59–57 opens, letting $12C$ become charged by electrons flowing up through $4P$ to 50. This is the heat time, as set by $4P$. When $12C$ has charged enough to bring grid 57 above point-2 potential, tube 6 fires;* electrons pass from 2 through tube 6 and $23R$ to point 60.

Just before tube 6 fires, notice that its anode 28 is at 60 potential, so that the right-hand side of $9C$ is also at this positive potential.

* The shield grid of tube 6 is connected at 38 to a negative potential; therefore tube 6 does not fire until its control grid becomes positive. Connected in the same way at point 37, tube 5 also has become a "positive control" tube; it is fired only by a positive control grid.

However the 52 side of $9C$ is at much lower potential (about 40 volts above point 2, this voltage being the combined arc drops of tubes 3, 4 and 5, all passing current). So the left side of $9C$ is about 230 volts more negative than the right side. Now, when tube 6 fires (at the end of heat time), its anode 28 drops instantly to within 15 volts of point 2; since the right side of $9C$ drops to this lower potential, the charge on $9C$ forces the left side 52 far below point 2. This makes the tube-5 anode more negative than its cathode, so the tube-5 current stops and weld relay $2CR$ drops out. Since the tube-5 control grid already is negative, as mentioned above, this grid regains control to prevent tube 5 from refiring when its anode 52 again becomes positive.

During the heat time, another $2CR$ contact (at lower right in Fig. $3D$) connects capacitor $28C$ through $25R$ to positive point 74 on $6P$; $28C$ becomes charged so that point 76 reaches a potential more positive than 2. Meanwhile the tube-7 control grid is held (through $57R$, $S2$ and $28R$) at the -75-volt potential of point 3. When $2CR$ drops out (at the end of heat time), its contact 72–76 connects the charged $28C$ into the grid circuit of tube 7; $28C$ forces electrons through $26R$, $S2$ and $28R$ so that grid 69 rises above point 2; tube 7 fires instantly, to start the hold time.

Before tube 7 fires, its anode 45 is near the same potential as point 50; capacitor $8C$ has practically no charge, and there is too little voltage across tube 8 for it to be able to fire. When tube 7 fires, point 45 drops to within 15 volts of point 2; grid 54 of tube 8 quickly drops to a potential about 50 volts above 2. More slowly, capacitor $8C$ becomes charged by electrons flowing from point 2 through tube 7, $34R$ and $2P$. After the hold time as set by $2P$, the potential of terminal 47 of $8C$ (which is also cathode of tube 8) will have dropped close enough to grid 54 so as to fire tube 8. When tube 8 fires, it discharges $8C$ through $P2T$. At bottom center, $S2T$ delivers a 100-volt impulse that raises the control grid of tube 2. Before tube 2 fires, notice that its anode 30 is at point-1 potential; meanwhile tube 1 is firing so its anode 22 is only 15 volts above 2; $6C$ is charged so its left-hand side 22 is more negative than point 30. When $S2T$ fires tube 2, anode 30 drops instantly to within 15 volts of point 2; the charge on $6C$ forces anode 22 of tube 1 far more negative than its cathode 2, so the tube-1 current stops, dropping out $1CR$. The $1CR$ contacts let the solenoid valve separate the electrodes; $1CR$ disconnects 60 from 1, so anode voltage is removed from tubes 4, 5, 6, 7, 8 and 10.

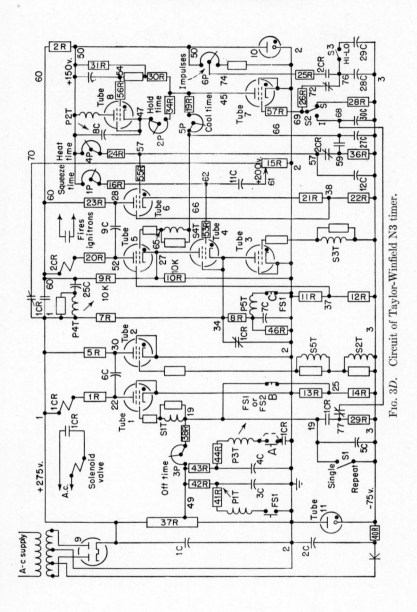

Fig. 3D. Circuit of Taylor-Winfield N3 timer.

Since switch $S1$ holds point 19 negative at 3, another weld cannot be started unless the foot switch has been released (to let $3C$ be recharged), and then closed again to refire tube 1.

3–6. Repeat Weld. Off Time. If switch $S1$ is opened for repeat (at lower left in Fig. $3D$), the $1CR$ contact 19–77 keeps point 19 negative during the squeeze, heat and hold times, but tube 1 continues to fire until its anode 22 is forced negative by tube 2 and $6C$, as described above. At once the negative potential at point 19 regains control of tube 1. However, when $1CR$ drops out at the end of hold time, capacitor $5C$ becomes charged by electrons flowing from its terminal 19 through $38R$ and $3P$ to 49. After this off time as set by $3P$, point 19 rises close to cathode 2 and refires tube 1 to start another welding operation.

3–7. Pulsation Weld. Cool Time. At the lower right in Fig. $3D$, throw switch $S2$ from S to I (for impulse welding). Now the tube-7 grid circuit includes $30C$; this capacitor has zero charge at the start of each welding operation (for $30C$ has been shorted by a $1CR$ contact at the left). During the first heat time, the $2CR$ contact lets $28C$ charge up to the voltage 3–74, as described above. Then, during cool time, the 72–76 contact of $2CR$ connects $28C$ through $26R$ to $30C$. Since $28C$ is much smaller than $30C$, the large voltage on $28C$ produces a much smaller voltage when applied across $30C$; this voltage is not large enough to fire tube 7. But $30C$ retains this charge during the second heat time while $28C$ again is being charged. During each cool time, the voltage across $30C$ is increased further until, after the desired number of impulses, its terminal 68 rises high enough to fire tube 7. This is like filling a large pail ($30C$) by repeatedly emptying a smaller pail ($28C$) into it. The number of such heat times or impulses may be increased by moving downward the slider 74 on selector $6P$. To obtain very few impulses, switch $S3$ connects $29C$ into circuit, thereby increasing the amount of charge added into $30C$ during each operation of $2CR$.

During each heat time, the $2CR$ contact 59–66 has discharged $27C$ through $36R$ (at lower right); terminal 66 drops to the negative potential of point 3. When tube 6 fires (turning off tube 5 by forcing its anode 52 negative), the tube-5 grid 65 is negative and regains control to hold off tube 5. At the end of heat time, $2CR$ drops out, letting $27C$ become charged by electrons flowing from $27C$ through $5P$, $19R$, $30R$ and $31R$ to positive point 50. After this cool time (whose length is set by $5P$), $27C$ terminal 66 has raised grid 65 to a potential high enough to fire tube 5, again picking up $2CR$ for another impulse of welding current. During the cool time, the 57–59 contact of $2CR$ has discharged $12C$ through $36R$, to be ready to measure the next heat time.

3–8. Two-stage Foot Switch. To use a two-stage foot switch, remove the $FS1$ n-c contact from B and connect it at C. Remove the dotted jumper at A; at A connect the second-stage contact $FS2$ (normally open) and place the normally-closed $FS2$ contact at B. Closing only $FS1$ (the first stage) fires tube 1 as before, picking up $1CR$ and bringing the electrodes together. But $1CR$ contact opens (2–34, at lower center of Fig. 3D), letting $7C$ charge to the voltage 1–2; $FS1$ is also open at C. Because of this charge on $7C$, cathode 34 of tube 4 is raised so high that tube 4 cannot fire. Likewise there is no $S3T$ voltage to fire tube 3. The weld relay $2CR$ is not picked up to fire the ignitrons. When the foot switch is released, the $FS1$ contact closes at C; $7C$ discharges through $P5T$ so that $S5T$ produces a 100-volt pulse to fire tube 2. Tube 2 turns off tube 1 as explained above, and $1CR$ separates the electrodes.

However, if both $FS1$ and the second stage $FS2$ are closed, $FS2$ and a $1CR$ contact complete the circuit at A so that $4C$ discharges through $P3T$. At lower center, $S3T$ produces a 100-volt peak to fire tube 3; the tube-3 arc drop holds cathode 34 of tube 4 to within 15 volts of point 2, so that tube 4 is fired normally after the squeeze time. The welding operation will be completed as described earlier.

3–9. Westinghouse Sequence Timer (N2 Combination). As is shown in Fig. 3E, this equipment includes a $3B$ sequence timer (having 10 tubes and 2 relays) enclosed with its ignitron contactor, thereby becoming a NEMA N2 combination. The diagram of this sequence timer appears in Fig. 3F. This circuit operates from 115 volts a.c. supplied at the left through a transformer (not shown). Notice that the top terminal 2 connects to the cathodes of thyratron tubes 1, 2, 5, 6 and 7; these tubes may fire only during those half cycles when bottom terminal 4 is more positive than terminal 2. Similarly, thyratron tubes 3, 4, 8, 9 and 10 may fire only during the opposite half cycles, when 2 is more positive than 4. A tube in one group may control a tube of the other group by the action of a capacitor.

For example, before starting to weld, tube 2 is not firing; therefore no electrons flow from top terminal 2 through tube 2 and

23R to terminal 4. Since there is no voltage across 23R, the control grid 19 of tube 3 is at cathode-4 potential and tube 3 fires. However, when the starting switch closes and fires tube 2 (as is described later), electrons flow through tube 2 and 23R so as to produce a voltage across 23R; 8C becomes charged by this voltage

FIG. 3E. NEMA N2 combination timer. (Westinghouse Electric Corporation.)

so that terminal 19 becomes more negative than terminal 4. A half cycle later, this charge on 8C keeps grid 19 negative so that tube 3 cannot fire. This same type of action is described earlier in Sec. 1–3.

Briefly, before the starting switch closes in Fig. 3F, tubes 3, 4, 7 and 8 are firing. Closing the starting switch at once fires tubes 1 and 2, picking up relay 1CR to let the solenoid valve bring the electrodes together; tube 2 turns off tubes 3 and 4. After the squeeze time, tubes 5 and 6 fire, picking up relay 2CR whose contacts fire the ignitron contactor (not shown). Tube 6 turns off tube 8; after the weld time, tube 9 fires, turning off tubes 5 and 7 and dropping out $2CR$ to stop the welding current. After the hold time, tube 10 fires and turns off tube 1, dropping out the solenoid valve so that the electrodes separate. If set for repeat operation, tube 2 also is turned off, again letting tubes 3 and 4 fire, turning off tube 6. Quickly tube 8 fires, turning off tube 9, thereby letting tube 7 fire. After the off time, if the starting switch still is closed, tubes 1 and 2 again fire, bringing the electrodes together for another welding operation.

3–10. Stand-by Conditions, before the Weld. Before the starting switch is closed, tubes 1 and 2 cannot fire, for their shield grids are held negative by the charge on capacitor $4C$. Tube 3 is firing (as described above) and electrons flow through tube 3, $3R$ and $2R$, also through $4R$ and $8R$; $4C$ is charged by the voltage across these resistors. Meanwhile the control grids of tubes 1 and 2 are near cathode potential; these grids connect through $10R$, $1P$ and $13R$ to cathode 2. At this time there is no charge on $19C$ and no voltage across $1P$ or $13R$. Across $10R$ there is a small a-c voltage produced by $S3T$; this voltage makes the control grids more positive at the start of those half cycles when the tube anodes are positive.

The transformer primary $P3T$ is shown at the lower left in Fig. 3F, where it receives voltage in the phase-shifting network $P2T$, $1R$, $1C$. Separate $S3T$ windings produce voltages across $10R$, $16R$ and $29R$; the amount of voltage across each resistor is set by a calibrating resistor C, which has been selected so as to give accurate time values to adjusters $1P$, $2P$ and $3P$.

Tube 4 is firing, so electrons flow through tube 4 and $2P$, charging capacitor $9C$ so that point 24 remains far more negative than cathode 2 of tubes 5 and 6. (The a-c voltage across $16R$ is smaller than the $9C$ voltage and does not affect tubes 5 and 6 at this time.) Since tubes 5 and 6 are not firing, no current flows through $45R$; there is no voltage across $36C$ so tube 8 has zero grid voltage and is firing. The anode voltage of tube 8 is supplied by transformer winding $S2T$; electrons flow from $S2T$ terminal 37 through $4P$ and tube 8, charging $32C$ so that point 37 is held more negative than cathode 4 of tube 9.* Since tube 9 is not firing, there is no

---

* The a-c voltage of $S2T$ is connected also across $8P$ and $47R$; part of this voltage is phase-shifted by $34C$ and $43R$ and provides means for calibrating weld-time adjuster $4P$.

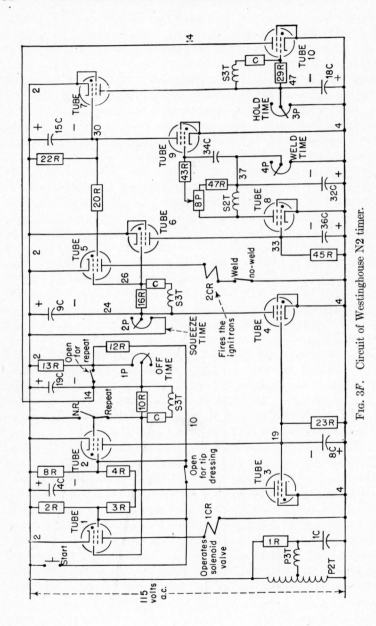

Fig. 3F. Circuit of Westinghouse N2 timer.

voltage across $22R$ or $15C$, so tube 7 has zero grid voltage and is firing. Its electrons flow through $3P$, charging $18C$ so that point 47 is held negative; tube 10 does not fire, so tube 10 causes no voltage across $13R$, $1P$ or $19C$.

3-11. The Weld Sequence. When the starting switch is closed in Fig. $3F$, the tube-1 shield grid is connected to cathode-2 potential and tube 1 fires, picking up relay $1CR$ so that the solenoid valve brings the electrodes together. (If the tip-dressing switch is open, tube 2 does not fire; the electrodes separate when the starting switch is opened.) With the tip-dressing switch closed, the starting switch also fires tube 2; electrons flow through tube 2 to charge $8C$, thereby holding negative the grids of tubes 3 and 4. Since tube 3 no longer charges $4C$, the shield grids of tubes 1 and 2 rise to cathode-2 potential so that these tubes continue to fire even if the starting switch is opened; this provides a nonbeat feature.

Tube 4 no longer charges $9C$; the voltage across $9C$ decreases at the rate set by squeeze-time adjuster $2P$. The point-24 potential approaches cathode 2 until the a-c voltage across $16R$ raises grid 26 high enough to fire tubes 5 and 6. If the weld–no-weld switch is closed, tube 5 picks up relay $2CR$ to fire the ignitron contactor and start the welding current. At the same time, electrons pass through tube 6 and $45R$; capacitor $36C$ becomes charged so that grid 33 is held negative, turning off tube 8.

This is the weld time, and $32C$ receives no further charge through tube 8; point 37 rises as $32C$ discharges through the weld-time adjuster $4P$. The rising potential at 37 lets the a-c voltage across $34C$ rise high enough to fire tube 9, to end the weld time. Electrons flow through tube 9 and $22R$, charging $15C$ so that its terminal 30 remains negative; this drives negative the shield grid of tube 5 so that tube 5 drops out relay $2CR$, stopping the welding current. Tube 7 also is turned off, so that $18C$ receives no further charge.

During the hold time, $18C$ discharges at a rate set by hold-time adjuster $3P$; point 47 rises high enough to permit the a-c voltage across $29R$ to fire tube 10. If the switch is set for non-repeat operation, electrons flow from cathode 4 through tube 10 to point 14, through the switch and $13R$ to point 2; capacitor $19C$ is charged to the voltage across $13R$, so that point 14 becomes negative, turning off tube 1 by its control-grid potential. Tube 1 drops out relay $1CR$ so that the electrodes separate. Since the control grid of tube 2 is connected to cathode 2, tube 2 continues to fire so long as the starting switch is held closed, for this keeps the tube-2 shield grid also at cathode potential. When the starting switch is opened, the tube-2 shield grid drops to the negative potential at point 14; the tube-2 current stops and $8C$ discharges quickly through $23R$, letting tubes 3 and 4 fire to reset the circuit for the next weld operation, in the following steps. Tube 3 charges $4C$ to hold negative the shield grids of tubes 1 and 2. Tube 4 charges $9C$; tube 6 is turned off, letting $36C$ discharge quickly through $45R$. Tube 8 fires, charging $32C$ and turning off tube 9. Capacitor $15C$ discharges quickly through $22R$, letting tube 7 fire and charge $18C$. Tube 10 is turned off and $19C$ discharges quickly through $13R$.

If the switch is set for repeat operation in Fig. $3F$, tube 10 is fired as before, at the end of the hold time. Electrons flow through tube 10 to point 14, through $1P$ and $13R$ to point 2. Point 14 becomes negative; the control grids turn off tubes 1 and 2 although the starting switch is yet closed. As described above, tubes 3 and 4 fire, turning off tube 6, letting tube 8 fire. Tube 9 is turned off, letting tube 7 fire; tube 10 is turned off. During the off time, capacitor $19C$ discharges through $13R$ and the off-time adjuster $1P$, letting the point-14 potential rise. If the starting switch still is held closed, the control grids again fire tubes 1 and 2 after the off time, to start another weld operation.

CHAPTER 4

G–E HIGH-SPEED SEQUENCE COMBINATIONS

For many years most sequence weld timers have included four to ten relays along with tubes in the time-delay circuits. Recent timers may retain only a few relays, while using additional tubes so that the timer is nearly all electronic with fewer moving parts. Examples of this trend are shown in Chap. 3 and in the description below.

4–1. Timer Arrangement (CR7503–L102S15). This is a NEMA type N2 welding control; as shown in Fig. 4A, it includes a standard ignitron contactor together with a sequence weld timer (NEMA 3B) which provides squeeze, weld, hold and off times. This timer has only two relays, $SVTD$ which controls the solenoid valve to bring the welding tips together onto the work, and WTD which fires the ignitron tubes. The timing circuits include no relays; thyratron tubes (GL502A or GL2050) are used instead.

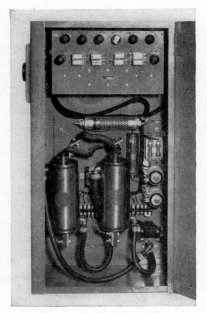

Fig. 4A. General Electric N2 combination timer.

At the right-hand side of the elementary diagram shown in Fig. 4H, a-c power supply is connected to ignitron tubes A and B, in series with the welding transformer. A WTD contact closes between the tube ignitors to fire these tubes. The a-c line voltage operates the heaters of the water-flow switch FS, and also picks up the welder solenoid valve when the $SVTD$ contact closes. Line voltage also passes through fuses to the primary winding $P1T$ of the control-power transformer; the main secondary winding of this transformer $1T$ is shown at the far left of the diagram ($S1T$). Other $S1T$ windings heat the tube filaments.

At the left, the center tap of $S1T$ is connected to line 10, which is at ground potential; line 10 connects to the cathodes of thyratron tubes 1 to 7, and is used as the reference line in this circuit. Between the two ends of this $S1T$ winding 140 volts a.c. appears. During a positive half cycle A (when terminal $70A$ is more positive than terminal $70B$), the upper diagram line $70A$ is 70 volts above (more positive than) cathode line 10; at the same time the lower diagram line $70B$ is 70 volts below (more negative than) cathode line 10. Since the anodes of tubes 1, 2 and 6 receive power from $70A$, these three tubes can fire only during A half cycles; similarly, tubes 3, 4, 5 and 7 can fire only during B half cycles, when $70B$ is more positive than cathode line 10. Before the starting switch closes, no thyratrons fire.

4–2. Operating Sequence. When starting switch $FS1$ closes, tubes 1 and 2 fire, in Fig. $4H$. Tube 1 causes $SVTD$ and the solenoid valve to bring the electrodes together. Tube 2 starts the squeeze timer, and also fires tube 3 so that the welding sequence will be completed even if the starting switch is released. After the squeeze time, tubes 4 and 5 fire. Tube 5 starts the weld timer; tube 4 makes relay WTD fire the ignitrons, to let welding current flow. After the weld time, tube 6 fires, starting the hold timer; tube 6 also turns off tube 4 (through transformer $6T$), and drops out WTD to stop the welding current. After the hold time, tube 7 fires. Through transformer $7T$, tube 7 turns off tube 1; relay $SVTD$ drops out and the welding tips separate.

If the switch is set for nonrepeat welding, tubes 2, 3, 5, 6 and 7 continue to fire until the starting switch is released. If set for repeat welding, the firing of tube 7 turns off both tubes 1 and 2, and the welding tips separate. As tube 2 stops, it turns off tubes 3, 5, 6 and 7. As tube 7 stops, the off timer starts. If the starting switch still is closed, tubes 1 and 2 again fire after the off time; the welding tips return onto the work, and the welding sequence is repeated.

4–3. Starting-switch Action. Before the starting switch closes, the control grids of tubes 1 and 2 (Fig. $4H$) are at zero or cathode potential, since tube 7 is not firing and there is no voltage across $71C$ or $P7T$ (at the right, above tube 7). However, tubes 1 and 2 are held off by their shield grids alone; these grids (nearest to anode) are connected through $26R$, $1C$ and $S7T$ to point 33 and through $3X$ to point $70B$. Since $S7T$ produces no voltage at this time, and since tube 3 is not firing or applying voltage across reactor $3X$, the shield grids of tubes 1 and 2 are at point-$70B$ potential. Therefore, during the half cycles A when the anodes of tubes 1 and 2 are positive, the shield grids are held negative at $70B$. During each half cycle B, when $S1T$ raises $70B$ and point 21 more positive than cathode 10, capacitor $1C$ is charged by electrons flowing from cathode 10 through tube 1 to grid 11 (or through tube 2 and $6R$ to 21) and through $26R$ to $1C$, then through $S7T$ and $3X$ to $70B$. This $1C$ charge helps to drive point 21 below cathode 10 before the start of half cycle A, thereby preventing unwanted firing of these tubes.

If a two-stage starting switch is used, the first stage permits bringing the welding tips together without completing the weld operation. When the switch connects grid 11 to cathode 10, tube 1 picks up $SVTD$; tube 2 does not fire. The welding tips again separate when the switch is released. However, when the second stage connects point 21 to 10, both tubes 1 and 2 fire, since all grids then are at zero or cathode potential.

To see how tube 2 (working in half cycle A) fires tube 3 (in half cycle B) we first must recall how a thyratron behaves with various kinds of load.

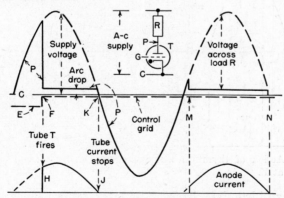

Fig. 4B. Action of thyratron with resistance load.

4–4. Thyratrons with Resistance Load or Inductive Load. In Fig. 4B, a thyratron T in series with load R is connected across the a-c supply voltage. So long as the tube grid G is so negative (as shown at E) that tube T does not fire, the entire supply voltage

58 ELECTRONIC MOTOR AND WELDER CONTROLS

appears between cathode C and anode or plate P. However, when the grid potential rises close to cathode C (as at F) and fires tube T at this instant, the potential at P drops at once to about 15 volts above C. Now the only voltage across the tube (from C to P) is the arc drop of the tube; the rest of the a-c supply voltage appears across the load. In Fig. 4B we assume that this load R is all resistance (or is noninductive). With resistance load, the anode current in tube T and load R rises abruptly at H, then follows a sine wave which reduces to zero at J, when the supply voltage also is zero. At K notice that anode P immediately becomes negative as the a-c supply voltage becomes negative. Since tube T acts as a rectifier, no current flows while P is below C. When P again rises above C in the next positive half cycle, the grid lets tube T fire at M; anode current flows again until the supply voltage reverses, at N.

Now notice the difference when the tube load is assumed to be inductive (such as a transformer winding or a reactor); this is the condition shown in Fig. 4C, where the load is reactor X. As before, so long as grid G prevents tube T from firing, the whole a-c supply voltage appears between anode P and cathode C. Also,

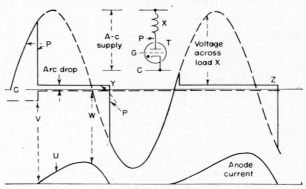

Fig. 4C. Action of thyratron with inductance load, showing "inductive hangover" at Y.

when the grid lets the tube fire (at V), the voltage across the tube again decreases suddenly until it is equal to the arc drop. However, notice that the anode current in tube T and load X rises more slowly (at U); moreover, due to the energy stored because of this current flowing in the inductive load X, the current does not decrease to zero at W, but continues to flow for some time after the supply voltage has reversed. So long as this tube current flows, the voltage across the tube remains equal to the arc drop; the potential at anode P remains positive far into the following half cycle, as shown at Y and at Z. This feature of "inductive hangover" may next be put to practical use.

4–5. Tube 2 Fires Tube 3. Figure $4D$ shows these two thyratrons as they appear in the complete sequence weld timer of Fig. $4H$. Whenever tube 2 fires (in half cycle A) it makes tube 3 fire in the following half cycle B.* As shown in Fig. $4E$, the a-c voltage supplied to tube 2 and $2X$ is positive during half cycles marked A. A half cycle later, when supply terminal $70A$ becomes negative, terminal $70B$ becomes positive so that tube 3 receives positive anode voltage during half cycles marked B. Therefore tubes 2 and 3 operate in alternate half cycles, much like a common two-tube rectifier. Since a negative grid keeps tube 2 from firing during the first A half cycle shown, the tube anode 23 is at the same potential as $70A$. Notice that anode 23 connects also to the grid of tube 3. During the next half cycle B, when tube 3 receives positive anode voltage from $70B$, the tube-3 grid has followed downward with the potential of $70A$. Such negative grid potential keeps tube 3 from firing.

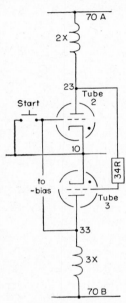

Fig. $4D$. By "inductive hangover," tube 2 fires tube 3.

The starting switch closes at E in Fig. $4E$, firing tube 2, so that current flows through tube 2 and its load $2X$. At once anode 23 drops to 15 volts above cathode 10, also holding the tube-3 grid at this positive potential. Since $2X$ is a reactor (an inductive load) the tube-2 current flows until F, far into the next half cycle B, as described above. Because of this "inductive hangover," the tube-3 grid is kept positive during the early portion of half cycle B. Since the tube-3 anode is positive in this half cycle B, tube 3 fires.

* In later circuits such as shown in Fig. $6C$ or Fig. $8B$, tube 2 may appear below the cathode center line while tube 3 is above it. In that case tube 2 (in half cycle B) fires tube 3 in the following half cycle A. The action is the same as described above.

Since tube 3 is a vapor-filled tube, it continues to fire for the rest of half cycle B even though its grid returns negative at G. So, for every half cycle that tube 2 fires, tube 3 fires the following half cycle.

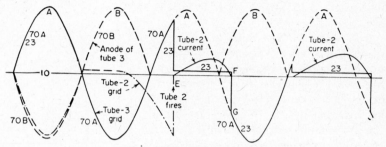

FIG. 4E. Wave shapes of action of tubes in Fig. 4D.

4–6. After a Time Delay, Tube 2 Fires Tube 4. Tubes 2 and 3 appear again in Fig. 4F, along with tubes 4 and 8A (as used in the weld timer of Fig. 4H). We will see that capacitor 21C provides the squeeze time of the weld timer by discharging through 21R at the desired rate until tube 4 is permitted to fire. This action is shown in Fig. 4G, where 21R is set for 3-cycle squeeze time.

Before tube 2 is fired by the starting switch, its anode 23 is at the same potential as 70A; so, during half cycle A, terminal 23 of capacitor 21C is at positive potential. Electrons now flow from cathode 10 up through tube 8A (which is used mainly as a diode rectifier) and into 21C; electrons continue from 21C through 2X to 70A. In this way 21C becomes charged to about 100 volts d.c., negative at terminal 26. In Fig. 4G, as the potential at terminal 23 starts to decrease at M, the charge on capacitor 21C forces terminal 26 more negative than cathode 10, as shown at N; this stops the electron flow through tube 8A. During half cycle B, while points 70A and 23 are negative, terminal 26 is yet more negative, although 21C is discharging slowly through 21R. Each half cycle A recharges 21C, as shown at Q.

When the starting switch closes, tube 2 fires (at R) during the half cycle A. Since anode 23 now remains 15 volts (arc drop) above cathode 10, capacitor 21C may continue to discharge through 21R and 22R, thus letting terminal 26 rise slowly toward cathode 10, as shown at S. Capacitor terminal 26 also controls the grid of tube 4. Since squeeze-time adjuster 21R is set for 3 cycles, terminal 26 will rise high enough, at U, so as to fire tube 4 in half cycle B (in the third cycle after the starting switch closes).

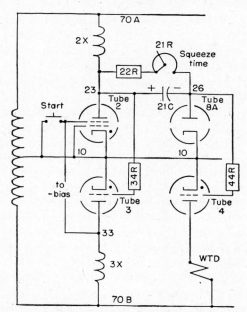

FIG. 4F. After squeeze-time delay, tube 2 fires tube 4.

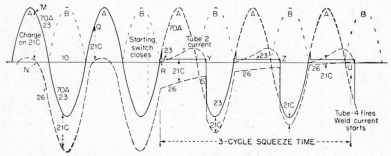

FIG. 4G. Wave shapes of time-delay action of tubes in Fig. 4F.

In Fig. 4F, we see that tube 8A lets 21C be charged during half cycles A, but disconnects 21C from 10 at all other times. When tube 2 fires, tube 3 fires the following half cycles (as at Y and Z, described above [4-5]), but tube 4 waits until a half cycle B at the end of the squeeze time. Tube 4 picks up relay WTD to fire the ignitrons. Now let us fit this action into the complete sequence weld timer.

4–7. Squeeze and Weld Times. Coming now to the complete diagram, Fig. $4H$, recall that tubes 1 and 2 are held off by the negative potential that $70B$ and $1C$ apply to the shield grids before the starting switch closes. The switch then fires tube 2, making tube 3 fire also; tube 3 applies voltage across $3X$ so that anode 33 is held about 15 volts more positive than cathode 10. Because of reactor $3X$, the tube-3 current continues to flow during the early portion of half cycle A, keeping point 33 positive so that the shield grids continue to fire tubes 1 and 2, even though the starting switch is quickly released. This provides the nonbeat feature whereby the starting switch may not interrupt the weld operation.

As described above, tube 2 also starts the squeeze time, letting $21C$ discharge through $21R$. At the end of squeeze time, point 26 and the control grid have risen high enough to fire tube 4. If pressure switch PS has closed,* WTD starts the weld current. While not shown in Fig. $4F$, point 26 also controls the grid of tube 5, which therefore fires at the same time as tube 4; tube 5 starts the weld time.

Before tube 5 fires, notice that capacitor $52C$ has been charged (much like $21C$, described earlier). During half cycle B, when points $70B$ and 53 are positive, electrons flow from 10 up through a low-voltage winding $S1T$† and through tube $8B$ to terminal 59 and into capacitor $52C$; electrons continue from terminal 53 through $5X$ to $70B$. When tube 5 fires, point 53 remains 15 volts

---

* Tube 4 cannot yet fire until its shield-grid circuit 52 is connected to line 10, since 52 is connected through $56R$ to 63 and the potential at $70A$. Some welding machines include a pressure switch whose contacts PS are connected into the circuit between 52 and 10. If no pressure switch is used, a jumper is placed from 52 to 10.

† This $S1T$ winding raises tube-$8B$ cathode 8 by about 10 volts during half cycle B. This makes sure that tube 8 passes no current until point 59 has risen close enough to cathode 10 so as to fire tube 6. Because of the energy in its emitted electrons, tube 8 may conduct slightly while its anode is a few volts more negative than its cathode.

Tube $8A$ passes no current during the early portion of any half cycle A; owing to the charge on $81C$ (received by grid rectification during a previous half cycle A), the tube-$8A$ grid remains negative during this early portion. In this way the charge on $21C$ is not affected by the higher voltage (greater than the arc drop) that exists across any thyratron just at the start of each conducting half cycle.

(arc drop) above line 10; the charge on capacitor 52C keeps terminal 59 negative, so that tube 8B no longer recharges 52C, but the control grid of tube 6 is held negative. After the length of weld time as set by 51R and the range switch 2S*, point 59 rises close to cathode 10 and fires tube 6 during a half cycle A. Tube 6 starts the hold time as next described; tube 6 also ends the weld time, by turning off tube 4. Tube 6 applies voltage to P6T, whose secondary S6T produces a voltage that drives negative the tube-4 shield grid. Although tube 4 is turned on by its control grid, its shield grid stops the tube-4 current, dropping out relay WTD to end the flow of welding current.

4–8. Hold Time. In Fig. 4H, tube 8A serves now for charging the hold-time capacitor 61C (above tube 6) as it served before for charging the squeeze-time capacitor 21C. During weld time, relay WTD not only fires the ignitrons, but its WTD contacts (shown above tube 8A) also switch the tube-8A anode circuit.† Since squeeze time has ended, WTD disconnects tube 8A from terminal 26 of 21C; another WTD contact connects tube 8A to terminal 65 of 61C. So, during weld time, electrons flow from 10 up through S1T and tube 8A to 85, through WTD contact to 61C, and from 61C terminal 63 through P6T to 70A.‡ When tube 6 fires, point 63 remains 15 volts (arc drop) above line 10; the charge on capacitor 61C keeps terminal 65 and tube-7 grid negative. After the length of hold time as set by 61R, point 65 rises close to cathode 10 and fires tube 7 during a half cycle B.

4–9. Nonrepeat Welding. If switch 1S is set for nonrepeat welding, 1S connects the control grid of tube 2 to cathode 10, so that tube 2 is controlled only by its shield grid 21. If the starting switch has been released, point 21 is being held positive at point 33

---

* Range switch 2S may select a weld time between 2 and 30 cycles or, in "High" position, between 30 and 120 cycles. In "High" position, 51C is added across 52C; these capacitors must discharge through 51R, 52R and 53R. In "Low" position, 51C and 53R are removed from this time-delay circuit, decreasing the time constant.

† In a recent circuit arrangement (not shown) the WTD contacts are not used above tube 8A; anode 85 is connected solidly to point 26, and 61C is charged by grid rectification.

‡ Before WTD picks up, 61C is charged by grid rectification, by electrons flowing from 10 down through P7T (at lower right in Fig. 4H), cathode 72 to control grid 74 of tube 7, to 61C; then through P6T to 70A. During squeeze time and weld time, tube 7 cannot fire, because its grid 74 is connected, past 61C, to the potential of 70A, which is negative whenever tube-7 anode is positive.

(see Sec. 4–7) so tube 2 continues to fire during the squeeze, weld and hold times. When tube 7 fires and applies voltage to $P7T$, a voltage is produced by $S7T$ (at the left) which drives point 21 negative and turns off tubes 1 and 2. However, so long as the starting switch is held closed, tube 2 continues to fire; tube 1 is turned off by its control grid 14 (as next described), so $SVTD$ drops out, releasing the solenoid valve so that the welding tips separate.

Before tube 7 fires, its cathode 72 is at line-10 potential, and $71C$ has no charge; terminal 14 (control grid of tube 1) also is at 10 potential and does not affect tube 1. At the end of hold time, tube 7 passes current through $P7T$; during half cycle B, electrons flow also from cathode 10 to the control grid 14 of tube 1 and into $71C$, then through tube 7 to $70B$. Capacitor $71C$ becomes charged, negative at 14. During the following half cycle A, the tube-7 current stops, letting cathode 72 drop back to point-10 potential; the new charge on $71C$ forces grid 14 far negative during this half cycle A, thereby preventing tube 1 from firing. So, even though the starting switch remains closed, the tube-1 current stops; $SVTD$ lets the welding tips separate but there is no further action. When the starting switch is released, the voltage produced by $S7T$ prevents tube 2 from firing, so its anode 23 returns to the potential of $70A$. In the next half cycle B, tube 3 does not fire; point 26 remains negative, so the tube-5 current stops, letting anode 53 return to the potential of $70B$. In the next half cycle A, point 59 remains negative; the tube-6 current stops, letting anode 63 return to $70A$. In the next half cycle B, point 65 remains negative; the tube-7 current stops, and capacitor $71C$ quickly loses its charge (since switch $1S$ shorts $71R$ during nonrepeat welding). All circuits have been reset, ready for the starting switch to be closed for the next weld.

4–10. Repeat Welding. When switch $1S$ is set for repeat welding, the tube-2 control grid is connected to tube-1 grid 14. At the end of hold time, tube 7 fires; the new charge on $71C$ now forces these grids negative; current stops in both tubes 1 and 2. Relay $SVTD$ drops out and the welding tips separate. In turn, current stops in tubes 3, 5, 6 and 7. After the tube-7 current stops, capacitor $71C$ discharges during this off time. Since switch $1S$ is not shorting $71R$, the length of off time is set by $71R$, through which $71C$ discharges. At the end of off time, grid 14 rises close to cathode 10; if the starting switch still is closed, tubes 1 and 2

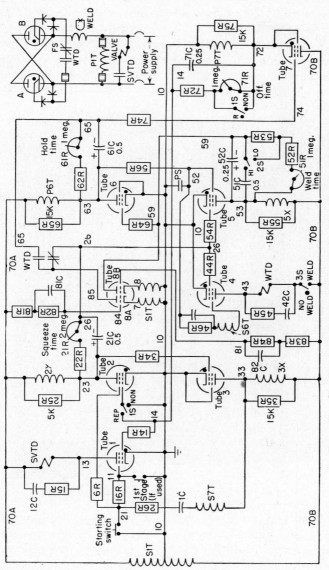

Fig. 4H. Circuit of combination timer (G-E type CR7503-L102S15).

fire, starting another welding operation. If the starting switch is open, tubes 1 and 2 are held off by the $70B$ potential applied to the shield grids at 21.

4–11. Automotive Welding Timer (CR7503–L127). A combination timer used in high-production industries is shown in Fig. $4I$; its diagram appears in Fig. $4J$. This timer provides the usual

FIG. $4I$. Automotive welding timer (CR7503–L127).

NEMA–3B sequence of squeeze, weld, hold and off, but has the added feature of double squeeze time (shown within the dashed enclosure in the center of Fig. $4J$ and described in Sec. 4–15). Six thyratron tubes determine the welding sequence, assisted by three 6H6 twin diode tubes. At the lower right, relay SVR energizes the solenoid valve; relay WR closes the ignitor circuit to fire ignitrons A and B to apply voltage to the welding transformer. The circuit is first described without reference to the double-squeeze circuit.

Briefly, the starting switch fires tubes 1 and 4, picking up the solenoid valve to bring together the electrodes. A half cycle later, tubes 2 and 3 energize their anode transformers $2T$ and $3T$; the $2T$ voltage prevents tube 7 from firing. At the end of the squeeze time tube 2 stops firing, thus removing the voltage of $2T$; the $3T$ voltage now fires tube 7, picking up relay WR and firing the ignitrons. Near the end of weld time, tube 3 stops firing; tube 7 drops out WR to end the flow of welding current. After the hold time, tube 4 stops firing; its relay SVR drops out the solenoid valve to separate the electrodes. If set for repeat welding, and if the starting switch still is closed, tubes 1 and 4 again are fired after the off time, to start another welding operation. This sequence of tube operation is sketched in Fig. 4K, to be explained later.

4–12. Circuit Conditions before the Weld. Power for welding and for the control transformer $1T$ is supplied at the lower right in Fig. 4J. At the upper left, $S1T$ supplies a total of 140 volts a.c., divided by a grounded center tap at point 10. Note that point 10 connects to the cathodes of tubes 2, 3, 4 and 7; while tubes 1, 4 and 7 receive anode voltage from $S1T$ terminal 170 so that they may fire during positive or A half cycles, tubes 2 and 3 receive anode voltage from the opposite terminal 270 during what we shall call negative or B half cycles of the power supply. (This arrangement of tubes and midtapped transformer is like that explained in Sec. 4–1 and shown in Fig. 4H.)

At the lower left in Fig. 4J, other secondary windings of transformer $1T$ supply voltage for the tube heaters (not shown); also, because of metallic rectifiers Rec 1 and Rec 2, and capacitors $1C$ and $2C$, these $1T$ windings produce d-c voltages that hold circuit point 5 at 20 volts below (more negative than) ground 10, and hold point 8 at -40 volts.

Before the starting switch closes (at the left), no tube has anode current, so transformers $2T$, $3T$ and $4T$ are not energized; there is no current through $3R$ so point 9 is at -40-volt potential. There is no charge on $12C$, so the control grid of tube 1 (connected through $15R$, $13R$, $14R$ and $3R$ or through $15R$ and $4R$) is also at the -40-volt potential of point 8; tube 1 is not firing, so its cathode 18 is at the -20-volt potential of point 5.

There is no charge on $22C$, so the tube-2 control grid connects

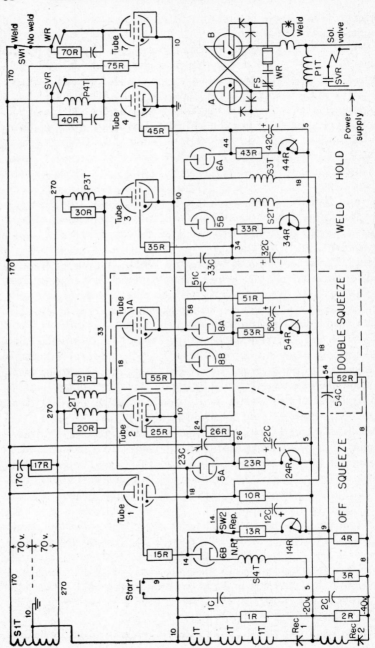

Fig. 4J. Circuit of automotive timer (CR7503–L127).

through 25R, 26R, 23R and 24R to the -20-volt potential at 5. Similarly, the tube-3 grid connects through 35R, 33R and 34R and the tube-4 grid connects through 45R, 43R and 44R, both grids being at -20 volts. The tube-7 grid connects through 75R, $S2T$ and $P3T$ to terminal 270 of $S1T$ (which is negative whenever the tube-7 anode is positive); tube 7 cannot fire so long as transformers 2T and 3T are not energized.

4–13. Starting the Weld. When the starting switch connects points 9 and 10, this brings point 14 and the tube-1 control grid to point-10 potential; since cathode 18 is 20 volts lower, tube 1 can fire when its anode next becomes positive, as is shown at E in Fig. 4K.* Electrons flow from point 5 through 10R and tube 1 to 170. Most of the 170-to-5 voltage appears across 10R so that its terminal 18 becomes about 110 volts more positive than 5. Electrons flow from 22C through tube 5A to 18, so that terminal 26 of 22C rises 110 volts above point 5, as shown at G. At the same instant, electrons flow from 42C through tube 6A and the $S3T$ winding to point 18 so that terminal 44 of 42C rises 110 volts above point 5. Since this rise at point 44 also lifts the grid of tube 4 higher than its cathode 10, tube 4 fires in this first half cycle with tube 1.

Tube 4 picks up relay SVR to bring the electrodes together, and also energizes transformer 4T. At the left in Fig. 4J, the voltage produced by $S4T$ makes electrons flow from terminal 9 of 12C up through $S4T$ and tube 6B to 12C terminal 14; this charges 12C so that terminal 14 and the tube-1 control grid are driven about 110 volts below ground 10, as shown at H in Fig. 4K. Thus tube 1 is prevented from firing more than a single A half cycle, just long enough to charge 22C and 42C. In the following B half cycle, when the anodes of tubes 2 and 3 become positive, the charge on 22C holds the tube-2 grid positive so that tube 2 fires, energizing transformer 2T. At the lower right in Fig. 4J, the voltage produced by $S2T$ forces electrons up through 34R, 33R and tube 5B, charging 32C so that its terminal 34 becomes more positive, raising also the grid of tube 3. At once tube 3 fires,

---

* If the starting switch closes midway in an A half cycle, as is shown at the left in Fig. 4K, tube 1 does not fire at once but waits until E, the beginning of the next A half cycle. The shield grid of tube 1 is very negative during each A half cycle except at its beginning. Because of 17R and 17C, the shield grid receives an a-c voltage that lags slightly behind the voltage wave of point 270, as is shown at F. This prevents tube 1 from firing until E.

energizing $P3T$ within this same B half cycle; this action occurs at J in Fig. 4K.

The stopping of tube 1 lets its cathode 18 drop at once to point-5 potential (for tube 5A disconnects 18 from the charge on 22C); however, the voltage of $S3T$ (at the lower right in Fig. 4J) renews the charge in 42C, by forcing electrons through 10R, 44R, 43R and tube 6A.

Since 22C receives no further charge through tube 1, the potential at its terminal 26 decreases as 22C discharges through 23R and squeeze-time adjuster 24R; thus, at K in Fig. 4K, the grid of tube 2 becomes more negative than its cathode 10 and prevents tube 2 from firing again.*

Since both tubes 2 and 3 are firing during the squeeze time, both transformers 2T and 3T produce voltage in the grid circuit of tube 7. The secondary voltage of 2T (shown above tube 2 in Fig. 4J) adds to the holdoff voltage at point 270; the voltage across $P3T$ is greater than the 2T voltage and opposes it, trying to turn on tube 7. So long as tube 2 is firing, the 2T voltage (combined with the 70-volt wave between points 270 and 10) holds the tube-7 grid negative. At the end of squeeze time, tube 2 stops firing and removes the 2T voltage; now the voltage of $P3T$ (in a B half cycle) fires tube 7 in the following A half cycle, at L in Fig. 4K, because of the inductive-hangover action described in Sec. 4–4. Tube 7 picks up relay WR, whose contacts (at lower right in Fig. 4J) close the circuit between the ignitors so as to fire the ignitron tubes A and B.

The removal of transformer-2T voltage not only lets 3T fire tube 7; since $S2T$ (at lower right) does not recharge 32C, this capacitor discharges at a rate set by 33R and weld-time adjuster 34R. At M in Fig. 4K point 34 and the grid of tube 3 become more negative than cathode 10;† tube 3 cannot fire during half cycle P, so tube 7 does not energize WR in half cycle Q. The

---

* When 24R is set for long squeeze time, 22C discharges so slowly that the tube-2 grid potential may happen to fire tube 2 midway in a half cycle, as is shown at W at the upper right in Fig. 4K. To prevent possible resulting error, a cyclic ripple is added to the grid potential, as shown by the dashed line; such a varying grid can fire tube 2 only at the beginning of its half cycle. This ripple is provided by capacitor 23C. A similar ripple is provided by 33C at the grid of tube 3; by 51C at the cathode of tube 1A.

† Note that anode current in thyratron tube 3 does not stop at M; current continues until after the anode voltage reverses, and thus fires tube 7 in half cycle N.

G-E HIGH-SPEED SEQUENCE COMBINATIONS 71

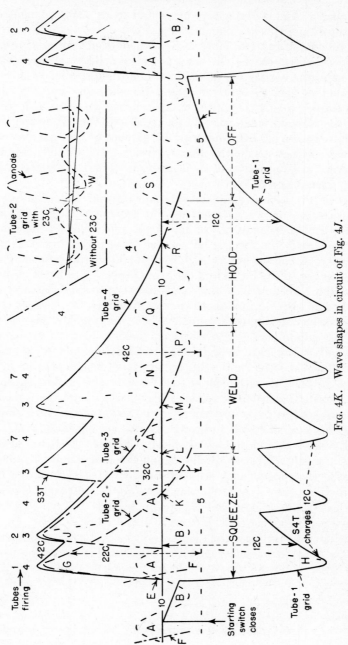

Fig. 4K. Wave shapes in circuit of Fig. 4J.

opening of the WR contacts prevents further firing of ignitrons A and B, and the welding current stops.

4–14. Hold and Off Times. Tube 3 does not fire during half cycle P in Fig. 4K, so $S3T$ does not recharge 42C; this capacitor discharges at a rate set by 43R and hold-time adjuster 44R. At R point 44 and the grid of tube 4 become more negative than cathode 10; tube 4 cannot fire in half cycle S, so relay SVR drops out the solenoid valve to separate the electrodes. If switch $SW2$ is set for nonrepeat welding, the tube-1 control grid remains connected (through 4R) to the negative potential of point 8, so tube 1 cannot start another weld sequence until the starting switch has been opened and reclosed.

If $SW2$ is set for repeat welding and the starting switch still is closed, tube 1 may start another weld sequence after the off time. When tube 4 does not fire (in half cycle S of Fig. 4K) there is no voltage from $S4T$ (at the left in Fig. 4J) to recharge 12C; this capacitor discharges at a rate set by 13R and off-time adjuster 14R. At T in Fig. 4K point 14 and the grid of tube 1 become more positive than its cathode, which is at point-5 potential; tube 1 fires again at U, the beginning of the next A half cycle.

4–15. Double Squeeze Time. When a gun welder is used for making a number of spot welds on a single piece of metal, the electrodes may need to be widely separated while getting into position before the first weld, but they need to separate only a small distance when they move to new locations to make additional welds on the same piece. To provide such action, the sequence timer is set for repeat operation; there should be a long or normal squeeze time before the first weld, but a shorter squeeze time for each following weld, so long as the starting switch remains closed.

At the center of Fig. 4J, the circuits including tubes 1A and 8 provide the squeeze-time adjustment (by 54R) for only the first weld operation; for the following welds, until the starting switch is released, the squeeze time is adjusted by 24R. For such action, 54R is set for longer squeeze time than the setting of 24R.

Before the starting switch closes, point 9 is at point-8 potential, so that there is no charge on 54C (bottom of Fig. 4J) and no voltage across 52R; the tube-1A grid is at point-8 potential, 20 volts below the tube-1A cathode (connected through 51R to point 5). Tube 1A is not firing; there is no voltage across 51R and no charge on 52C.

Just as the starting switch raises point 9 to point-10 potential, capacitor $54C$ raises point 54 also, thereby firing tube $1A$. However, after a few cycles $54C$ has charged (through $52R$) to the voltage between points 8 and 10, so point 54 returns to point-8 potential, thus holding negative the grid of tube $1A$; thus tube $1A$ fires just as the starting switch closes, but does not fire during the repeat welding operations thereafter, so long as the starting switch remains closed.

When tube $1A$ fires, electrons flow from point 5 through $51R$ and tube $1A$ to point 18 and through tube 1 to 170. Because of the voltage across $51R$, electrons flow from 5 through $54R$, $53R$ and tube $8A$, charging $52C$ to nearly the same voltage (18-to-5) that charges $22C$. When tube 1 is kept from firing after its first half cycle,[4–13] this also stops the current flow through tube $1A$ and $51R$; thus $52C$ is left to discharge through $54R$, while $22C$ discharges through $24R$. Since $24R$ is set for shorter time so that $22C$ discharges more quickly, the potential at its terminal 26 becomes negative faster than the potential at terminal 51 of $52C$. Therefore electrons flow from point 26 through $26R$ and tube $8B$, to point 51, thereby holding the grid of tube 2 at the higher potential of point 51.

Each time the starting switch is closed, tube 2 fires during a squeeze time as adjusted by $54R$. If the starting switch is kept closed, for repeat welding, tube $1A$ does not fire again (as explained above) when tube 1 fires to recharge $22C$. The voltage across $52C$ has decreased nearly to zero, so point 51 has lower potential than point 26; tube $8B$ prevents current flow between these points, so the tube-2 grid responds only to the lowering potential at 26. Thus the squeeze time of following welds depends on the setting of $24R$, until the starting switch is released; then $54C$ discharges (through $52R$ and $3R$) to be ready for the next series of welds.

CHAPTER 5

SYNCHRONOUS TIMING

Although the sequence weld timers of earlier chapters may control the flow of welding current for times as short as two or three cycles, greater accuracy is needed for welding certain metals and for producing the same amount of heat in each spot weld.

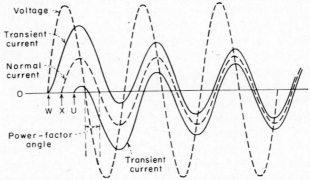

Fig. 5A. Transient currents (starting before or after the power-factor angle).

5-1. Synchronous Timing. To get these results, an a-c welder must be controlled by tube circuits that can start the flow of welding current always at the same point in the 60-cycle voltage wave. Figure 5A shows this voltage wave; it pictures the current that flows if the weld is started at points W, X or U. Since a welding transformer is a lagging-power-factor load, the welder tries to draw current that lags (starting at X) behind the voltage; if the welding current begins to flow at X, its wave shape remains the same, one cycle after another, and gives the best a-c weld. However, if the contactor closes the welding circuit at W (perhaps only $\frac{1}{1000}$ sec earlier than X), the amount of current rises much higher than before; for 3 or 4 cycles this current is "off balance" before it returns to normal. This short-time disturbance is called a *transient current*. Similarly, if it is started late at U, the current again is much greater (below the 0 line). Whether started early or late, this increased current changes the amount of heat during that weld. So, especially when the welding current flows for less than five cycles, far better welds are made when tube-operated circuits start the flow of current always at point X. Such accurate starting of the welds, always exactly in step with the a-c voltage wave, is called *synchronous timing*.

A small synchronous weld timer is shown in Fig. 5B; its elementary diagram appears in Fig. 5C. Here we shall see that tube 7 starts the flow of welding current always at the same point in the voltage wave. Similarly, the heat-control circuit of Sec. 1–5 also starts the current to flow at carefully controlled points in the voltage wave, thereby providing one of the desired features of synchronous timing.

Most synchronous spot-welding timers include a circuit that lets the welding current flow for 1, or 2, or 10 cycles, or other lengths of time, as selected by turning a dial. These timers also make sure that each weld includes an even number of half cycles of current flow—that is, the current flows for 2 or 5 complete cycles, but not for 1½ or 4½ cycles.* If the weld starts with a positive half cycle, it always ends with the opposite, or negative, half cycle. In this way, a new weld never starts with the same kind of half cycle that ended the previous weld.†

As shown in Fig. 5C, tubes 7 and 8 are connected back to back to permit alternating current to flow to the welding transformer. Tube 7 always is the first to pass current—tube 8 always is the last to pass current. The part of the synchronous timer that causes this action is called the *leading-tube–trailing-tube circuit*.

5–2. Synchronous Control for Small Welders (CR7503–A140).

For welders needing less than 50 amperes from the a-c supply line a pair of thyratrons may be used instead of ignitrons to switch this current flow. In Fig. 5C thyratron tubes 7 and 8 are used in this way. Briefly, when the starting switch closes, this fires tube 2. Tube 3 fires also and, through transformer 4T, turns off tubes 5 and 6 and lets transformer 2T fire tubes 7 and 8, to make the weld. After the desired weld time set by 35R, tube 4 fires, canceling the effect of tube 3 and stopping tubes 7 and 8.

---

* Some small timers may include a dial setting that causes a half-cycle weld.

† This is important because the iron in the welding transformer retains some magnetism from that final half cycle; this will affect the current and the heat of the next weld if the current is permitted to start in the same direction (or start with the same kind of half cycle that ended the previous weld).

76   ELECTRONIC MOTOR AND WELDER CONTROLS

Fig. 5B. Synchronous control for a small spot welder (CR7503–A140).

SYNCHRONOUS TIMING

The a-c power enters at the bottom of Fig. 5C; the main current flows through tubes 7 and 8, the welding transformer and the contacts of relay PCR. The a-c supply also connects to the proper voltage tap of transformer 1T at upper left, thereby energizing P3T and filament transformers (not shown). After heating the tubes 5 minutes, relay PCR may be pushed in to make ready to weld. If the weld–no-weld switch 1S is closed, relay PCR is held closed through its own contact; if supply voltage fails, PCR drops out, giving undervoltage protection. With PCR closed, notice that point 5 (upper right) is at the same potential as point 2 (upper left).

The part of the P1T (primary of transformer 1T) winding between points 2 and 6 has the center tap 8 that connects to the cathodes of tubes 1, 2, 3 and 4. This arrangement of transformer and tubes is much like that shown in Fig. 4F and Secs. 4–5 and 4–6; as there explained, we must consider separately each half cycle of the supply voltage. Let us call half cycle A any half cycle when points 2 and 5 are more positive than midtap 8; these are the only half cycles when tube 3 or tube 4 can fire. (Tube 7 fires only during half cycles A.) Similarly, call the other half cycles B, when point 6 is more positive than midtap 8; these are the only half cycles when tubes 1 and 2 can fire. (Tube 8 fires only during half cycles B.)

5–3. Stand-by Grid Voltages. While the starting switch is open (at upper center of Fig. 5C), transformer 7T shifts the a-c grid voltage* of tube 2 so that it becomes negative (as shown at D at upper left in Fig. 5D) just before the tube-2 anode becomes positive in half cycle B; tube 2 cannot fire.

The grid 29 of tube 3 is connected through 29R to point 20; similarly, tube-4 grid 39 is connected through 39R, 35R, 32P and 31R to point 20. Since tube 2 is not passing current, to cause a voltage across reactor 1X, point 20 is at the same potential as point 6; as shown at E in Fig. 5D, this potential is out of phase

* If transformer 7T happens to have exactly the same reactance as capacitor 17C, their combined effect (resonance) would be similar to a resistor; the voltage between points 8 and 18 would be in phase with A (voltage 8-to-2) and therefore would be 180 degrees out of phase with B (voltage 8-to-6), which is the anode voltage of tube 2. Transformer 7T is so designed that, with open secondary, its reactance is greater than that of 17C, thereby advancing the phase of voltage 8–18, as shown at D in Fig. 5D. When the starting switch short-circuits its secondary, 7T has less reactance than 17C and the phase of voltage 8–18 lags, as at F, so that the tube-2 grid remains positive for a short time after the tube-2 anode becomes positive; this fires tube 2.

with half cycle A, which is the anode voltage of tubes 3 and 4. With such out-of-phase grid potential at point 20, tubes 3 and 4 will not fire. Since these tubes apply no voltage to $P4T$, tubes 5 and 6 prevent tubes 7 and 8 from firing, as explained later.

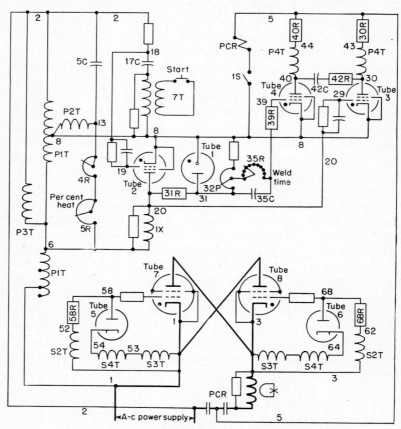

FIG. 5C. Circuit of synchronous control (CR7503–A140).

Meanwhile during half cycle B, when point 20 is more positive than midtap 8, electrons flow through tube 1, from cathode 8 to anode 31, and through $31R$ and $1X$ to point 6. Tube 1 is a voltage-regulator tube,[*] designed to hold 105 volts across itself, 8 to 31. During each half cycle B, point 20 rises to a voltage high enough to fire tube 1; any excess above 105 volts then appears across $31R$.

[*] See footnote, Sec. 2–4.

SYNCHRONOUS TIMING

During this same half cycle B, electrons flow also through tube 4 from cathode 8 to grid 39, and through $39R$, $35R$ and $32P$, $31R$ and $1X$ to point 6. These electrons charge capacitor $35C$; during the following half cycle $35C$ cannot discharge by this same path, because the tube-4 grid acts as a rectifier. Although the a-c supply voltage may vary, $35C$ always charges to the same voltage, thanks to voltage-regulator tube 1.

5-4. Starting the Weld Operation. When the starting switch closes in Fig. $5C$, the 18-to-8 voltage shifts to a new position (shown at F in Fig. $5D$) that keeps the tube-2 grid 19 positive for a short time after the start of half cycle B; this fires tube 2. At once tube-2 anode 20 drops to 15 volts (arc drop) above cathode 8, preventing further charging of capacitor $35C$. The glow in tube 1 goes out, for this tube passes no current when its 31–8 voltage is so low. Shown at G, the charge remaining on $35C$ keeps grid 39 so negative that tube 4 cannot fire until $35C$ has had time to discharge through $35R$, as described in Sec. 5–5.

During each half cycle in which tube 2 fires, point 20 and grid 29 of tube 3 both remain 15 volts above 8, not only during half cycle B but also far enough into the following half cycle A to fire tube 3. Thus tube 3 is fired by the "inductive hangover" from tube 2 and its inductive load $1X$ (see Secs. 4–4 and 4–5). So long as the starting switch is closed, tube 2 fires during each half cycle B, making tube 3 fire during each half cycle A. As shown later, tube 3 causes tubes 7 and 8 to let welding current flow, but this action stops when tube 4 fires.

When tube 3 is firing, voltage is applied to transformer $4T$ ($P4T$ winding 43 to 30); no voltage is applied to the other $P4T$ winding 44 to 40, since tube 4 has not yet fired. This condition lets $S4T$ produce voltage during half cycle A, to turn off tube 5 and fire tube 7 as is described in Sec. 5–6. However, during the following half cycle B when tube 3 cannot fire to apply voltage to $P4T$, we still need output voltage from $S4T$ to control tube 6 and let tube 8 fire. To obtain this missing half wave of $S4T$ voltage, $42C$ and $42R$ are added above tubes 3 and 4, between anodes 30 and 40.

As shown in (b) of Fig. $5D$, point 40 follows close to point-5 potential during the half cycle A when tube 3 fires, but the tube-3 arc drop holds point 30 close to cathode-8 potential. This 5-to-30 voltage is applied not only to $P4T$ (43–30); it also charges $42C$. At I the charge on $42C$ becomes zero; then $42C$ charges in the reverse direction. Since transformer $P4T$ is an inductive load,[4-4] the tube-3 current continues to flow (later than I) during the early part of the following half cycle B and leaves $42C$ charged (+) at 30, (−) at 40, as shown at H. When tube-3 current stops, $42C$

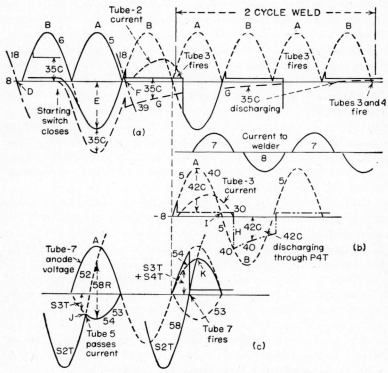

Fig. 5D. Wave shapes of circuit action in Fig. 5C.

then discharges during the remainder of half cycle B; its electrons flow from 40 up through $P4T$ and $40R$ to point 5, down through $30R$, $P4T$ and $42R$. This electron flow energizes $P4T$ during half cycle B so that $S4T$ provides the voltage needed to let tube 8 fire. This gives a leading-tube–trailing-tube action; tube 7 fires first to start the welding current, and always is followed by tube 8. Thus, when tube 3 fires for a half cycle, this lets both tubes 7 and 8 fire to cause a full cycle of welding current.

5-5. Weld Time. Earlier we mentioned that, when tube 2 fired, capacitor $35C$ began to discharge through $35R$. This $35R$ may be turned to select the desired number of cycles of weld time.

In the two-cycle position, the resistance in $35R$ will keep $35C$ charged long enough (as shown in Fig. $5D$) so that grid 39 cannot rise high enough to fire tube 4, until the start of the third cycle.* So, during the 2 cycles of weld time, tube 3 is firing but tube 4 is not firing. When tube 4 fires at the start of cycle 3, the tube-4 current energizes $P4T$ winding 44–40 at the same time that tube 3 energizes $P4T$ winding 43–30. These two $P4T$ windings oppose each other, so that $S4T$ produces no voltage. Therefore, when tube 4 fires, tubes 7 and 8 cannot fire. As next explained, tubes 7 and 8 can fire only when the $S4T$ windings produce voltage.

5–6. Making the Weld. Connected between points 2 and 6 (at upper left in Fig. $5C$) we find the heat-control phase-shifting network made up of $5C$, $4R$ and $5R$ in series, receiving the a-c voltage across $P1T$. Between their junction 13 and midtap 8 is transformer $2T$, whose primary voltage is phase-shifted by turning $4R$ or $5R$; this circuit is discussed in Sec. 1–5. The secondaries $S2T$ supply sine waves (240 volts) to the control grids of thyratrons 7 and 8; therefore these tubes may be phase-shifted by turning $5R$, to produce the desired heat at the weld.

Although $S2T$ always produces its voltage (as between points 1 and 52, at the lower left in Fig. $5C$), point 58 may rise high enough to fire tube 7 only when tube 5 is not passing current. Tube 5 is a simple rectifier; it will pass current so long as its cathode is more negative than its anode. The tube-5 anode 58 connects through $58R$ to $S2T$; the tube-5 cathode 54 depends on the voltages produced by $S3T$ and $S4T$. As described above, there is no $S4T$ voltage until tube 3 fires. At all times $S3T$ produces about 60 volts, and is connected so that it drives its terminal 53 negative during half cycle A, when the tube-7 anode is positive; this is shown in (c) of Fig. $5D$.

Before the starting switch closes, $S3T$ alone keeps cathode 54 of tube 5 more negative than point 1 during each half cycle A. When the $S2T$ voltage tries to raise point 58 higher than cathode 54 (as at J in Fig. $5D$), tube 5 passes current and keeps 58 close to the negative potential of 54. Since this keeps the control grid of tube 7 far more negative than its cathode 1, tube 7 cannot fire.

When the starting switch closes and tube 3 fires, $S4T$ (below tube 5) produces about 180 volts, opposed to the $S3T$ voltage. Since the $S4T$ voltage is larger than the $S3T$ voltage, cathode 54

---

* The effect of weld-time calibration $32P$ is not shown in Fig. $5D$; it is discussed in Sec. 6–5.

of tube 5 now is made more positive than point 1 (as at K in Fig. 5D). Tube 5 does not conduct until after $S2T$ already has raised point 58 high enough to fire tube 7. So, when tube 3 fires and $S4T$ prevents tube 5 from holding point 58 at negative potential, the voltage wave of $S2T$ fires tube 7; the firing point of tube 7 in half cycle A may be adjusted by per cent-heat dial 5R. Similarly, in half cycle B the other $S4T$ winding raises the cathode 64 of tube 6 so that $S2T$ (3 to 62) may fire tube 8.

CHAPTER 6

G–E SYNCHRONOUS COMBINATIONS

Since earlier designs of the synchronous weld timer have been described,* here we introduce more recent types of the synchronous timer, whose electric circuits and mechanical arrangement warrant attention. As shown in Fig. 1A and Fig. 6A the cabinet may be mounted directly as part of the welding machine; the single enclosure may hold also the sequence timer, thereby becoming a synchronous combination control.

6–1. One Synchronous Control for Many Uses. Equipments described in this and following chapters often control the larger sizes of welding machines, especially those that weld critical metals such as aluminum or stainless steel; here greater value is placed on the features of synchronous starting, the precise timing of the welding current, the ensured equal number of positive and negative half cycles of current flow.

Many varieties of synchronous combination are described—for a single spot weld, for a pulsation weld, for temper welding with post heat, or for welding with forge pressure. These types differ only in the kind of sequence control used in the upper portion of the enclosure; all use the same design of ignitron contactor, shown at the bottom in Fig. 6B.† The center portion includes the synchronous-timing, heat-control and firing circuits; these remain the same for any of the varieties mentioned above. Slightly modified, these circuits are used also in seam-welding controls (Chap. 8).

A complete synchronous spot-weld combination will be described in detail; then different sequence controls will be substituted to produce desired welding features.

6–2. Spot-weld Combination (CR7503–A151). This synchronous combination, shown in Fig. 6B, includes a simple sequence

---

* Chute, "Electronic Control . . .," Part II.
† For a motor-driven welding machine, the sequence panel may be omitted; a cam closes a contact to fire the central and lower sections.

control (NEMA 7B). The upper panel measures the squeeze, hold and off times; the weld time is controlled entirely by the synchronous-timer center panel. Figure 6D shows the diagram of the sequence control; three wires connect this to the synchronous-timing circuits of Fig. 6E. In Fig. 6E, tubes 1, 2, 3 and 4 perform like those in Fig. 5C, previously described. Similarly,

FIG. 6A. A spot welder with synchronous control.

in Fig. 6D, tubes 21, 25, 26, 27 and 28 correspond to tubes 1, 5, 6, 7 and 8 in Fig. 2E, as described in Sec. 2–4. Tubes 27 and 28 furnish a regulated d-c voltage so that upper point 75 is at $+60$ volts above grounded point 10; bottom point 70 is at -90 volts below 10.

In brief, when the starting switch closes in Fig. 6D, relay $1CR$ picks up the solenoid valve to bring the welder electrodes together. After the squeeze time, tube 21 picks up relay $1TD$. Near the starting switch, $1TD$ contacts fire tube 2 in Fig. $6E$. Tube 3 fires also and, through transformer $4T$, turns off tubes $6A$ and $6B$ and lets transformer $2T$ fire tubes 7 and 8; these tubes (through transformer $3T$) fire thyratrons 17 and 18 so that ignitrons 19

Fig. 6B. Synchronous spot-weld combination (CR7503–A151).

and 20 pass current to the welding transformer W. After the number of cycles of weld time set by $35R$, tube $5B$ fires tube 4, canceling the effect of tube 3, and thereby shutting off tubes 7, 8, 17, 18, 19 and 20. The per cent heat obtained during each cycle of welding current is selected by $5R$ (lower in Fig. $6E$).

At the end of the weld time tube 2 stops, turning off tube 22 in Fig. 6D; this starts the hold time. After the hold time, tube 25 picks up 5TD. One 5TD contact drops out 1CR to separate the welder electrodes; another 5TD contact (below tube 22) starts the off time. If 3S is closed (for a repeat weld) tube 26 fires after the off time, dropping out 5TD; 1CR again operates the solenoid valve to start another welding operation.

6–3. Stand-by Conditions, before the Weld. Let us study the synchronous-timing circuits in detail. The a-c power, supplied at the bottom of Fig. 6E, passes through a circuit breaker and ignitrons 19 and 20 to apply voltage to the welding transformer W. A control switch connects power to 1T and other transformers (including 21T which supplies the sequence control of Fig. 6D). After allowing 5 minutes for warming the tubes, time-delay relay TD closes the circuit to 7T, whose secondary S7T supplies voltages at the upper left in Fig. 6E. Notice that S7T is arranged like the anode transformer of a two-tube rectifier, for it has center tap G (grounded) and end taps 140A and 140B.* Let us call half cycle A any half cycle when point 140A is more positive than midtap G; this is the only half cycle when tubes 3, 4, 5B, 7 or 17 can fire. Similarly, call the other half cycle B, when points 140B and 70B are more positive than midtap G; this is the only half cycle when tubes 1, 2, 5A, 8 or 18 can pass current.

Before the starting switch closes (in Fig. 6D), a 1TD contact connects terminal 91 of capacitor 9C to the +60 volt potential of point 75; the other 1TD contact is open between 91 and the control grid of tube 2. No thyratron tube in Fig. 6E is firing. The tube-2 control grid 24 is connected through 19R and P4T to the a-c voltage at 140A; this voltage is far negative (as is shown at E in Fig. 6G) during each half cycle B when the tube-2 anode is positive (connected through 8T to 140B). The voltage across 19C (charged by tube-2 grid current in half cycle A) drives grid 24 negative before the start of half cycle B, as shown at D in Fig. 6G.

While tube 2 is not firing, its anode 25 is at the potential of 140B; this a-c voltage at 140B is applied also to the control grid of tube 3, keeping this grid negative during the entire half cycle while anode 33 is positive. Tube 3 cannot fire to apply voltage to P4T; neither can tube 4 fire, as next explained. Since there is

* The voltage between 140A and 140B is 280 volts rms, when measured by an a-c voltmeter. From 70A to G the voltmeter reads 70 volts, and 35 volts from 35A to G.

no $P4T$ voltage, tubes $6A$ and $6B$ pass current and prevent tubes 7 and 8 from firing, as explained in Sec. 6–6. Keep in mind that ignitrons 19 and 20 are fired only when the $S4T$ transformer windings produce voltage at the cathodes of tubes $6A$ and $6B$.

Meanwhile, during half cycle B when point 25 is more positive than center line G, electrons flow through tube $5A$, through $35R$, $32P$, $31R$ and $8T$ to point $140B$. These electrons charge capacitor $35C$. Tube $5A$ acts as a rectifier to prevent $35C$ from discharging during the following half cycle.* Since the a-c voltage supplied through $S7T$ may change due to line variations, the voltage-regulator tube 1 is added, so that $35C$ charges always to the same amount of voltage. (Later, when tube 2 fires, $35C$ will control the length of weld time as it discharges through $35R$.) During the A half cycles, when points 25 and $140B$ are negative but the anodes of tubes $5B$ and 4 are positive, grid $53A$ is driven negative (since rectifier $5A$ disconnects point $53A$ from G), so tube $5B$ passes no current. With no electron flow through cathode resistor $37R$, grid $52B$ of tube 4 is at the negative potential of $35B$, so tube 4 cannot fire.

* Tube $5A$ is grid controlled, to prevent it from passing current during the early part of the half cycle. As shown in Fig. $6C$, $46C$ has been charged during earlier half cycles B (by electrons flowing from G, cathode to tube-$5A$ grid,

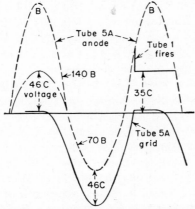

Fig. $6C$. Delayed firing of tube $5A$ (see footnote).

through $46R$ and $47R$ to $70B$). Retaining most of this charge one cycle later, $46C$ holds the tube-$5A$ grid negative until after the tube-$5A$ anode voltage at $140B$ has risen to about half its full amount. In this way $35C$ receives no charging current during the early part of the voltage wave, when VR tube 1 is not yet firing. See also footnote in Sec. 2–4.

88 ELECTRONIC MOTOR AND WELDER CONTROLS

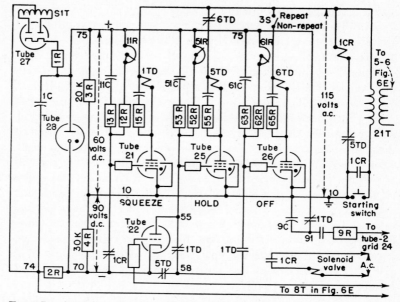

FIG. 6D. Sequence-timer circuit used with Fig. 6E to form a CR7503–A151 combination.

In Fig. 6D, tube 22 is not firing. Its grid is connected (through the secondary of transformer 8T in Fig. 6E) to the more negative potential at point 74.* Until tube 2 fires, 8T produces no voltage. Meanwhile the n-c contacts of 1TD and 5TD are closed around tube 22, to hold negative the tube-25 grid.

6–4. Starting the Weld. When the starting switch closes in Fig. 6D, the sequence control brings the welder electrodes together; after the squeeze time, relay 1TD operates its contacts. Terminal 91 of 9C now is disconnected from the +60-volt potential of point 75; this positive potential (held by the charge on 9C) is applied by TD1 contact, to raise the tube-2 control grid, as is shown at the upper left in Fig. 6G. However, this may not instantly fire tube 2, since the tube-2 shield grid is very negative in each B half cycle except at its beginning. Because of 17R and 17C, the shield grid of tube 2 receives an a-c voltage that lags

---

* Resistor 2R serves as ballast for VR tube 28; the voltage across 2R acts also as a negative bias to hold off tube 22.

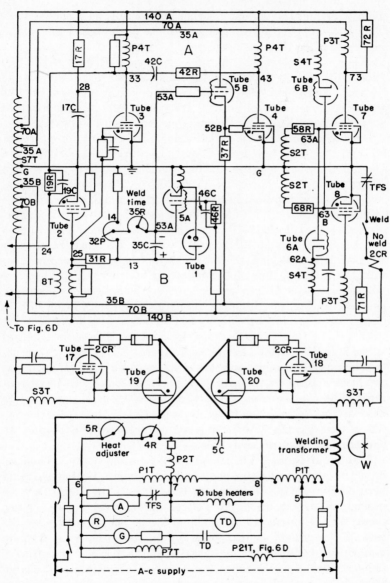

Fig. 6E. Synchronous timing circuit, controlled by circuit of Fig. 6D or Fig. 6H.

slightly behind the voltage wave of point $140A$, as is shown at F. This prevents tube 2 from firing until the beginning of the next B half cycle after $1TD$ has operated.

When tube 2 fires (in Fig. $6E$), electrons flow from G through tube 2 and $8T$ to $140B$; the tube-2 anode 25 drops to 15 volts above G, and thereby it (1) turns on tube 3, (2) turns on tube 22 in the sequence control and (3) starts the weld-time action that soon fires tube 4. These three actions are described next.

Since transformer $8T$ is an inductive load, current flows through $8T$ and tube 2 in half cycle B and also during the early part of the following half cycle A, holding point 25 positive long enough to fire tube 3, as is shown at Q in Fig. $6G$. (This "inductive-hangover" action is discussed in Sec. 4–5.) During half cycle A, tube 3 applies voltage to $P4T$ to let tubes 7 and 8 and the ignitrons fire, as is described in Sec. 6–6. But notice that anode 33 of tube 3 connects back through $19R$ to grid 24 of tube 2. Since transformer $4T$ is an inductive load, the tube-3 current holds anode 33 positive into the early portion of the following half cycle B. Therefore whenever tube 3 fires, it refires tube 2. This becomes necessary since $9C$ (in Fig. $6D$) becomes discharged after once firing tube 2; thereafter tube 2 is controlled by its grid potential at point 33. So, until tube 4 fires (to end the weld time, as is described below) tubes 2 and 3 continue to fire each other.

When tube 2 fires and applies voltage across $8T$, this transformer produces voltage at the grid of tube 22 to make it pass current during those half cycles when the tube-25 anode is positive. In this way, while tube 2 is firing during the weld time, tube 22 acts like a closed contact to hold point 55 negative and thereby to prevent the start of hold time.

6–5. Weld Time. When tube 2 first fires and point 25 becomes only 15 volts (arc drop) above G, this prevents further charging of capacitor $35C$. The glow in tube 1 goes out, for this tube passes no current when the voltage across it is so low. Since point 13 also is only 15 volts above G, the charge on $35C$ drives grid $53A$ negative, so that tube $5B$ passes no current; with no voltage across $37R$, the tube-4 grid is at the negative potential of point $35B$ so tube 4 cannot fire. Now $35C$ begins to discharge through $35R$, which may be turned to select the desired number of cycles of weld time. If in the two-cycle position, the resistance in $35R$ will keep $35C$ charged long enough so that grid $53A$ cannot rise high enough to let tube $5B$ fire tube 4 until the start of the third cycle of weld time (as is shown at T in Fig. 6G).* When tube 4 fires, it ends the weld and turns off tube 2, as described later.

So, during the 2 cycles of weld time, tube 3 is firing but tube 4 is not firing. When tube 3 alone applies voltage to $P4T$ (140A to 33 in Fig. 6E), but no voltage is applied to the other $P4T$ winding (140A to 43), this condition lets the $S4T$ winding produce voltage. As is shown later, this turns off tube 6B to let tube 7 be fired during half cycle A. However, during the following half cycle B, when tube 3 cannot fire or apply voltage to $P4T$, we still need output voltage from $S4T$ to control tube 6A and to let tube 8 fire. To obtain this missing half wave of $S4T$ voltage, 42C and 42R are added above tubes 3 and 4, between anodes 33 and 43. Now, each time tube 3 fires, point 33 is close to G potential; this voltage is applied not only to $P4T$ (140A to 33) but it also charges 42C, as shown in (b) of Fig. 6G. Since $P4T$ is an inductive load and the tube-3 current continues to flow during the early part of the next half cycle B, the final stopping of tube-3 current

* In Fig. 6G, the voltage across 35C is shown decreasing smoothly, as if the weld-time-calibration adjuster 32P were set so that its slider 15 touches at point 13. This action is shown again by the solid line in Fig. 6F, but for the condition where 35R is set for a longer weld time such as 20 cycles. If 32P now is turned toward point 14, part of the voltage 13-to-G is included in the discharge path of 35C. During half cycles B (with tube 2 firing and

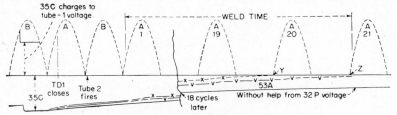

Fig. 6F. Weld-time calibration by 32P in Fig. 6E.

holding points 25 and 13 close to G) the voltage across 32P is very small. But during each half cycle A, the voltage across 32P is large; slider 15 is more positive than 13. This voltage has such polarity as to help 35C to discharge more rapidly. In Fig. 6F, line VVV shows how a medium setting of 32P will let 35C discharge faster, to fire tube 4 at Z. Suppose this produces a weld time of 21 cycles while 35R is set in the 20-cycle position. To correct or calibrate this setting, 32P is turned closer to 14, increasing slightly the voltage between slider 15 and point 13. This lets 35C discharge more quickly as shown by line XXX in Fig. 6F. Tube 4 is fired one cycle earlier at Y, to agree with the 20-cycle setting of 35R.

leaves 42C charged (+) at 33, (−) at 43, as shown at H. Capacitor 42C then discharges during the remainder of half cycle B; its electrons flow from 42C through 42R, from 43 up through P4T to point 140A, down through the other P4T winding to 33 and 42C. This electron flow energizes P4T during half cycle B so

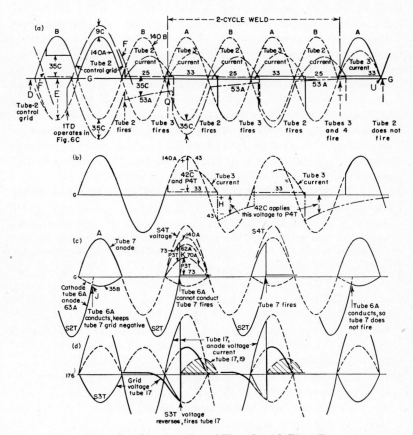

FIG. 6G. Circuit action of Fig. 6D with Fig. 6E.

that S4T provides the missing half wave of voltage needed to let tube 8 fire. This gives a leading-tube–trailing-tube action; tubes 7 and 17 fire first so that ignitron 19 starts the welding current. These always are followed by tubes 8 and 18 firing ignitron 20. Thus, the firing of a single tube 3 for a half cycle lets both tubes 7 and 8 fire, to cause a full cycle of welding current.

When tube 4 fires, at the end of the weld time, the tube-4 current energizes $P4T$ winding $140A$–43 at the same time that tube 3 energizes $P4T$ winding $140A$–33. These two $P4T$ windings oppose each other, so that $S4T$ produces no voltage. Therefore, when tube 4 fires, the effect is the same as if neither tube 3 nor tube 4 is firing; tubes 7 and 8 cannot fire, and the welding current stops. Also, when the two opposing $P4T$ windings are both energized, this cancels or removes the inductive part of $P4T$ as a load, so the tube-3 current now flows in phase with the supply voltage $140A$-to-G. Since the tube-3 current now stops at U [shown at upper right in (a) of Fig. 6G] and does not keep point 33 positive into half cycle B, the grid 24 of tube 2 does not remain positive in half cycle B, so tube 2 does not fire. Its anode 25 returns to the potential of $140B$, so the grids of tubes 3, $5B$ and 4 all are kept negative in the A half cycle that follows; all these tubes stop together. Capacitor $35C$ recharges to be ready for the next weld time.

6–6. Making the Weld. Now let us see how the $S4T$ voltages control tubes $6A$ and $6B$ so as to fire tubes 7 and 8 during the weld time. If you remove tube 6 (which consists of two diodes $6A$ and $6B$ in one enclosure), the voltages of $S2T$ then can fire tubes 7 and 8 every half cycle; this lets ignitrons 19 and 20 fire steadily.

These $S2T$ windings supply sine waves at 160 volts. The primary $P2T$ (in the lower part of Fig. 6E) is in the heat-control phase-shifting network made by $5C$, $4R$ and $5R$, connected in series across the 230 volts a.c. of $P1T$. By turning dial $5R$, the $P2T$ and $S2T$ voltages may be phase-shifted so as to control or change the point (within each half cycle of supply voltage) when tubes 7 and 8 are to be fired (see Sec. 1–5).

Now return tube 6 to the circuit. Either diode $6A$ or $6B$ will pass current whenever its cathode is more negative than its anode. With no voltage from either $S4T$ winding, notice that cathode $62A$ of tube $6A$ is at the potential of $35B$, which is negative during half cycle A, as shown in (c) of Fig. 6G. Now, when the $S2T$ voltage is rising toward G, it cannot raise point $63A$ much above $35B$, for tube $6A$ passes current at J; these electrons flow from $35B$ to $62A$ and through tube $6A$ and $58R$ to $S2T$. The $S2T$ voltage is wasted across $58R$, and grid $63A$ cannot rise high enough to fire tube 7. (During the B half cycle, tube $6B$ similarly passes current to hold grid $63B$ more negative than G, so that tube 8 cannot fire.)

When tube 3 fires during weld time, each $S4T$ winding produces a large voltage. The winding shown below tube $6A$ gives a voltage that opposes the small $35B$-to-G voltage and raises cathode $62A$ high above G. Now tube $6A$ cannot conduct until K; grid $63A$ already has risen high enough to fire tube 7. The $S4T$ voltages have blocked out tube 6 so that $S2T$ can fire tubes 7 and 8. When tube 4 fires and stops the $S4T$ voltage, tube 6 again passes current (as at J) and keeps grids $63A$ and $63B$ so negative that tubes 7 and 8 cannot fire.

6–7. The Welding Power Circuit. In the lower part of Fig. $6E$, thyratrons 17 and 18 fire ignitrons 19 and 20. Tube 17 is controlled entirely by tube 7 above and the voltage of transformer $3T$.

When tube 7 is not firing, the $P3T$ winding shown above it is receiving (through $72R$) the a-c voltage supplied between points $70A$ and $140A$ of $S7T$. The resulting $S3T$ wave (in the grid circuit of tube 17) is out of phase with the tube-17 anode voltage [as shown in (d) of Fig. $6G$] and prevents tube 17 from firing.

When tube 7 fires during the weld time, its anode 73 is only 15 volts above G; most of the voltage $140A$-to-G now appears across $72R$. $P3T$ now receives most of the $S7T$ voltage supplied between $70A$ and G, but this is exactly opposite in phase to the $70A$-to-$140A$ voltage it received previously. At the instant when tube 7 fires, the $S3T$ voltage reverses (in the tube-17 grid circuit) and becomes positive, in phase with the anode voltage, so as to fire tube 17. Since tube 7 is not fired by $S2T$ until part way through the half cycle, $P3T$ is not connected to the $70A$-to-G voltage until this same point, which is set by heat adjuster $5R$. As shown in (d) of Fig. $6G$, the grid voltage of tube 17 is kept negative during the first part of the anode-voltage wave, but the grid suddenly is switched to a positive potential at the instant when tube 7 fires. Exactly one-half cycle later, $S2T$ fires tube 8 to cause a similar reversal in the tube-18 grid circuit, to fire ignitron 20.

6–8. Hold Time, Off Time. When the tube-2 current stops at the end of weld time, $8T$ no longer produces voltage to fire tube 22 in the sequence control (Fig. $6D$). Point 55 now rises as $51C$ discharges at the rate set by hold-time adjuster $51R$. After the hold time, tube 5 picks up $5TD$; a $5TD$ contact drops out $1CR$ to separate the electrodes. Another $5TD$ contact starts the off time. After the off time (as set by $61R$), if $3S$ is closed to produce a repeat weld, tube 6 picks up $6TD$; a $6TD$ contact resets $5TD$.

If the starting switch still is closed, $1CR$ again is energized, to start another weld operation.

6–9. Pulsation-weld Combination (CR7503–A152). This equipment is like that just described except that a NEMA 9B sequence timer (upper portion) is used, to provide either a spot or a pulsation weld. As shown in the diagram, Fig. $6H$, this sequence timer is similar to that of Fig. $6D$ but tube circuits are added for cool time and weld interval. The length of each flow of welding current (heat time) is controlled by the synchronous timer as before. In this combination control, the center and lower sections (synchronous timer and ignitron contactor) are the same as described before and as shown in Fig. $6E$. Figure $6H$ shows the pulsation-weld sequence timer and its connections to the synchronous timer, part of which is shown at the right. By moving jumpers, several connections have been changed; the tube-2 shield grid is now connected to cathode. Resistor $17R$ is connected to $140B$; the junction 143 (of $17R$ to $17C$) is applied to the shield grid of tube 23 in the sequence timer; the tube-23 anode connects back to 33, the tube-3 anode at the right.

In brief, when the starting switch closes in Fig. $6H$, relay $1CR$ picks up the solenoid valve to bring the electrodes together. After the squeeze time, tube 21 picks up relay $1TD$. Near tube 21, $1TD$ contacts let tube 23 fire, so that $P4T$ (at the right) starts the welding current; tube 2 fires in the next half cycle. Tube 3 then fires in place of tube 23 to apply voltage to $P4T$. As shown previously in Fig. $6E$, the $S4T$ voltages keep tube 6 from passing current, so tubes 7 and 8 are permitted to fire; through $3T$, tubes 17 and 18 are fired, thereby firing ignitrons 19 and 20, to permit the flow of welding current. After the number of cycles of weld time as set by $35R$, tube $5B$ fires tube 4, canceling the effect of tube 3, and thereby shutting off tubes 7, 8, 17, 18, 19 and 20.

6–10. Tube-23 Circuit. In detail, before the starting switch closes in Fig. $6H$, the n-c $1TD$ contact holds the control grid of tube 23 at the -90-volt d-c potential of point 70. Since tubes 23 and 3 are not firing, their anode potential (connected at point 33 through $P4T$) is the same as at $140A$. As in the previous equipment (see Sec. 6–3), this a-c voltage at $140A$ is applied also through $19R$ to the control grid of tube 2. Tube 2 cannot fire, so tube 3 does not fire. No voltage is applied to transformer $8T$, so grid 24 of tube 22 is at the negative potential of point 74 (at lower left in Fig. $6H$). Neither half of tube 22 passes current (nor is relay

96 ELECTRONIC MOTOR AND WELDER CONTROLS

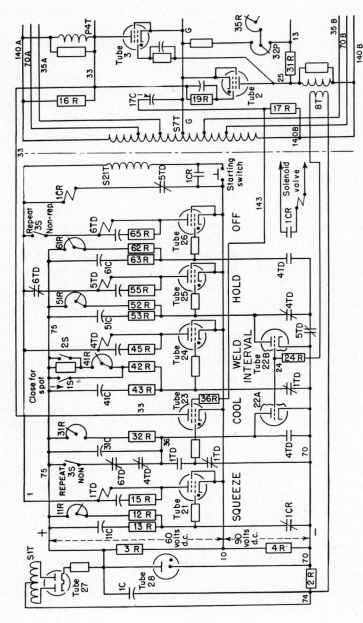

Fig. 6H. Pulsation-timer circuit used with Fig. 6E to form a CR7503-A152 combination.

$4TD$ yet picked up) so there is no voltage across $31C$; point 36 is at the $+60$-volt potential of point 75.

When the starting switch has closed and the electrodes are together, relay $1TD$ operates after the squeeze time. The $1TD$ contacts suddenly change the tube-23 control grid to the high potential at 36. This is shown at the left in Fig. 6I. This may not instantly fire tube 23, since its shield grid is very negative in each A half cycle, except at its beginning. Because of $17R$ and $17C$, the shield grid of tube 23 receives an a-c voltage that lags slightly behind the voltage wave of point $140B$, as shown at F in Fig. 6I. This prevents tube 23 from firing until the beginning of the next A half cycle after $1TD$ has operated.

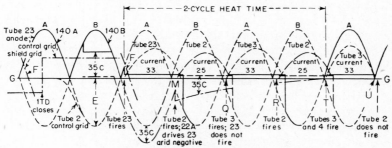

Fig. 6I. Circuit action of Fig. 6H with Fig. 6E.

When tube 23 fires, it applies voltage to $P4T$ (at the right in Fig. 6H); as described in Sec. 6-6, the $S4T$ voltages now control tubes 6, 7 and 8 (shown in Fig. 6E) so that welding current may flow. The firing of tube 23 also holds its anode 33 positive into the following half cycle B (as shown at M in Fig. 6I) so that tube 2 fires. Electrons flow from G through tube 2 and $8T$ to $140B$; the inductive load $8T$ holds tube-2 anode 25 positive into the following A half cycle (at Q in Fig. 6I), firing tube 3. Tube 2 also makes $8T$ produce a voltage at the grid of tube 22 so that current flows through tube $22A$, lowering point 36 and turning off tube 23. Although the tube-23 current stops, tube 3 now fires in place of tube 23 to keep $P4T$ energized. The "inductive hangover"[4-4] caused by $P4T$ lets tube 3 refire tube 2 (at R) even though tube 23 is no longer firing. In this way, tubes 2 and 3 will continue to refire each other until tube 4 fires, as explained later.

When tube 2 first fires, this starts the heat time,* to measure the number of cycles of welding-current flow. The synchronous-timer circuit now performs the complete operation described in Sec. 6–5.†

6–11. Cool Time. Weld Interval. When tube 2 is thus stopped at the end of the heat time, voltage is removed from $8T$, so the secondary of $8T$ does not force positive the grid of tube $22A$ in Fig. $6H$. This grid 24 returns negative to the bias across $2R$; the tube-$22A$ current stops and disconnects capacitor $31C$ from negative point 70. As $31C$ is now permitted to discharge through $31R$, the tube-23 control-grid potential gradually rises. After this cool time as set by $31R$, tube 23 may fire again (but only at the start of a half cycle A, because of its shield-grid voltage), and start another pulse of welding current. This alternate heat-cool action continues throughout the pulsation weld.

Meanwhile the number of welding-current pulses is controlled in the tube-24 grid circuit. At the end of squeeze time, a $1TD$ contact opens (below tube $22A$ in Fig. $6H$) letting the tube-24 grid potential rise slowly as $41C$ discharges at a rate set by the weld-interval adjuster $41R$ (and by $A41R$, if range switch $2S$ is open).‡ When tube 4 fires, to end the pulsation weld, one $4TD$ contact closes (below tube $22A$) to prevent tube 23 from firing again. Another $4TD$ contact closes (below tube 25) to turn off

---

* As outlined in Sec. 2–3, a pulsation weld is made by letting the welding current flow more than once, with a cool time between these periods of flow. The length of each flow is called the heat time; this corresponds to the weld time of an ordinary spot weld. The over-all sum of the heat and cool times is controlled by the weld-interval timer.

In the post-heat or temper-weld timer of Sec. 7–5, welding current flows first during the weld time. Then follows a period of no current flow, called the chill time. Current flows again, during the temper time, as controlled by the synchronous-timer weld-time circuit.

† In the timing sequence of the A152 pulsation timer, notice in Fig. $6I$ that the welding current starts in the half cycle A before tube 2 fires. In the A151 spot timer, (a) of Fig. $6G$ shows that the welding current starts in the half cycle A after tube 2 fires. In both controls, the time-delay action (of $35C$ discharging through $35R$) starts when tube 2 first fires. Therefore the A151 combination uses a resistor in series with $35R$ to delay the firing of tube 4 by one extra cycle of time; this extra resistor is shorted by a jumper added to the A152 combination.

‡ If switch $1S$ is closed shorting $41R$, $41C$ discharges so quickly that tube 4 picks up $4TD$ to prevent more than a single pulse of welding current; an ordinary spot weld results.

tube 6 and drop out 6*TD*. Above tube 21, contacts of 4*TD* and 6*TD* now discharge 31*C* (if 3*S* is closed for a repeat operation) to be ready for the next pulsation weld. When tube-2 current stops at the end of the next heat time, 8*T* removes its voltage from grid 24 of tube 22*B*. Tube-22*B* current stops, and the hold time starts as described in Sec. 6–7.

CHAPTER 7

SLOPE CONTROL, TEMPER WELDING AND FORGE TIMING

Often certain metals or parts cannot be welded satisfactorily by a simple spot or pulsation weld. Better welds may be obtained when accessory circuits are added to the timers already described; three such accessories are included here. The slope control pro-

FIG. 7A. Slope-control accessory (CR7503–D202).

vides gradual increase of current during each spot weld. The temper weld is made by several impulses of current; the first impulse makes the weld, which is then annealed by a second impulse (but with different heat and time values). A forge timer applies increased pressure at the welding tips at any desired time during the weld.

7-1. Slope Control (CR7503–D202). It is well known that aluminum and similar metals may be welded more satisfactorily (so as to require less frequent cleaning of the welding tips) if the welding current starts at a small amount and then rises to full welding value. Various welding-control systems already may include this feature of gradual current rise during the weld.*

A simple method of controlling this current rise is offered by slope control, an accessory that may be added to welding controls that include phase-shift heat control. Figure 7A shows a slope control. By its use, during each weld the a-c welding current is made to increase at a gradual rate (or slope) that may be adjusted to give best welding results.

The addition of slope control removes the flash often seen when heavy sections are being spot- or projection-welded; therefore, the pressure on the electrodes may be decreased. As a result, the line current or kva required for making a heavy weld may be decreased perhaps 50 per cent. Moreover, the decreased current during the first cycle tends to penetrate or correct the surface condition of the parts being welded; the more constant surface thus provided may permit more consistent welds than those obtainable by other simple welding controls.

Figure 7B shows first the current wave during an ordinary four-cycle weld with single-phase control. During this ordinary

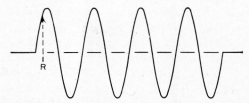

Fig. 7B. Usual welding-transformer current wave.

weld the amount of current may increase too quickly (as at R in Fig. 7B) before the electrode tips have made contact equally over all parts of the weld surface. With slope control added, Fig. 7C shows the small current during the first cycle (as at S); this softens the weld metal and lets the tips make better contact. Then the amount of welding current rises gradually to the full amount to complete the weld (at T).

The slope control provides this gradual current rise during each spot weld, acting like a variable heat adjuster in the welding

* These include the stored-energy systems (condenser-discharge and electromagnetic) popular during World War II; also the "three-phase" or frequency-changer systems (see Chap. 11).

102   ELECTRONIC MOTOR AND WELDER CONTROLS

equipment to which it is added. Figure 7D outlines how a slope control may be used with the phase-shift heat-control unit described in Sec. 1–5 and Fig. 1J; perhaps the welder is controlled

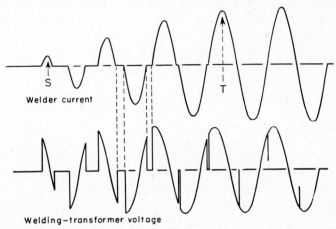

FIG. 7C. Welding-transformer current and voltage, using slope control.

by a NEMA 3B timer (see Sec. 2–1). Whenever the timer closes its contact to let thyratrons 1 and 2 (of the heat-control unit) fire the ignitrons, a relay XCR operates to start the action in the slope-control unit.* The whole slope-control circuit is shown in Fig. 7E.

In Fig. 7D notice that the slope control includes a pair of thyratrons X and Y, added in series with the heat-adjusting dial $5R$ of the phase-shifting circuit. At the start of each weld, tubes X and Y are firing very little; during the weld they are turned on gradually until they are firing completely during the last cycles of the weld. Since X and Y are connected back to back, a.c. may flow through them when their grids permit. When firing completely, tubes X and Y act like a short circuit, so the highest weld heat is adjusted by $5R$.† However, when tubes X and Y are not firing, the resistance of $2P$ is added to $5R$; this decreases the weld heat. Therefore $2P$ is used to adjust the starting heat—the

* In many sequence timers or combination controls, there may be an unused contact that closes as the welding current starts; this contact may be used for contact C in Fig. 7E, to pick up relay XCR. In circuits that use no relay to start the welding current, an extra tube circuit may be added, as outlined in Sec. 7–3.

† The weld heat is increased if the resistance in $5R$ is decreased, as when $5R$ is turned clockwise.

amount of current flowing in the first cycle of each weld; 5R adjusts the current flowing in the final cycles of that same weld.* In each weld, tubes X and Y gradually increase the welding current from the 2P setting to the 5R setting.

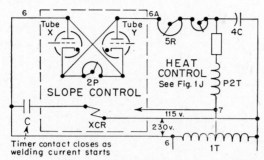

Fig. 7D. Slope control added to heat-control circuit.

7-2. Slope-control Circuit. At the top in Fig. 7E, the slope control receives 115-volt and 230-volt power from transformer terminals 6, 7 and 8 in the phase-shift circuit; P1T warms the filaments of tubes X and Y. With selector switch 1SW turned on,† tubes X and Y receive a-c anode voltage from the phase-shifting circuit; they do not fire until relay XCR operates. But the circuit now is complete from 6 through 1SW to 13, through 2P and 3R to 23 and 6A, to 5R of the heat-control circuit.

The grid circuit of tube X is like that of tube Y. Each grid circuit includes two transformer windings; each winding supplies perhaps 50 volts a.c. These are secondary windings of transformer 2T, whose primary (P2T at the upper right in Fig. 7E) is part of the fixed phase-shifting network that includes 1R and

---

* Dial 2P is marked "Heat Start—Per Cent Final Current." When 5R is set for 80 per cent final heat and 2P is set at 50 per cent, the starting current is 0.50 × 0.80, or 40, per cent of full, or final, current. If 5R now is turned to 60 per cent but 2P remains at 50 per cent, this setting gives 0.50 × 0.60, or 30, per cent at the start. To prevent erratic firing of the ignitrons, the 2P and 5R dials must be set so that the starting current is not less than 40 per cent of full current at 230 volts, nor less than 20 per cent at 460 volts.

† If 1SW is in the off position, tubes X and Y cannot fire; in place of these tubes, 1SW connects resistor 4R into the heat-control circuit. Resistor 4R causes about the same voltage difference between 6 and 6A as exists when tubes X and Y are in circuit, firing completely. Without this feature, only one ignitron might fire when the heat-control dial is set near 100 per cent (in a phase-shift circuit that uses a peaking transformer).

104 ELECTRONIC MOTOR AND WELDER CONTROLS

1C.* In the tube-X grid circuit, winding 4 of $S2T$ furnishes an a-c voltage wave that lags about 90 degrees behind the tube-X anode voltage. Starting at the tube-X grid, and tracing through 11R, $S2T$ and 12R to cathode 23, we see that this $S2T$ voltage wave adds to the voltage across capacitor 12C, to control tube X.

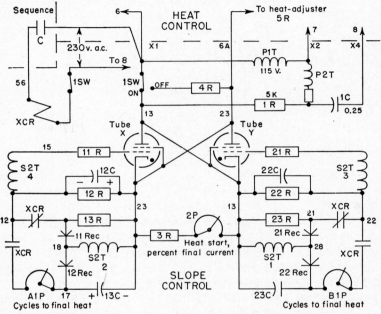

Fig. 7E. Circuit of slope control (CR7503–D202).

Before relay XCR operates at the start of each weld, the voltage of $S2T$ (winding 2) charges capacitors 12C and 13C. During a half cycle when $S2T$ terminal 18 is more positive than 23, the $S2T$ voltage charges 13C [(−) at 23]; electrons may flow from 13C terminal 17 through selenium rectifier 12 Rec to $S2T$. During the following half cycle 12 Rec prevents the reverse flow of electrons, so 13C remains fully charged (as is shown in Fig. 7F); this 13C voltage has no effect on tube X at this time. Mean-

* Notice that 1R, 1C and $P2T$ of the slope control are connected to terminals 6, 7 and 8 in the same way as 5R, 4C and $P2T$ are connected in the heat-control circuit. Since 1R is 5000 ohms and 1C is 0.25 mu f (equal to 10,640 ohms in a 60-cycle circuit) the vector triangle shows that the $S2T$ voltages (in the slope control) lag about 60 degrees behind the line voltage. So, at high-heat settings of 5R, $S2T$ lags perhaps 90 degrees behind the tube-X anode voltage.

while, when *S2T* terminal 23 is more positive than 18, electrons may flow through selenium rectifier 11 Rec, through the n-c contact of *XCR* and through 12*R* to 23; the voltage across 12*R* charges 12*C* so that its terminal 12 is negative.* This 12*C* voltage, added to the a-c voltage of *S2T* (winding 4) produces the wave of grid voltage shown at *L—M* in Fig. 7*G*; this voltage holds negative the tube-*X* control grid during each entire half cycle when the tube-*X* anode is positive. (In the same way, the tube-*Y* grid is held negative by 22*C* and *S2T* winding 3.) So long as *XCR* does not operate, tubes *X* and *Y* cannot fire.

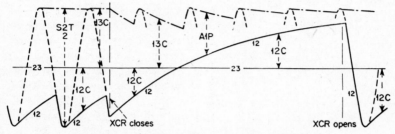

Fig. 7*F*. Charge on 12*C* changes when relay *XCR* is picked up.

When the sequence timer picks up relay *XCR*, its n-c contact disconnects point 12 from *S2T* (winding 2) so that 12*C* receives no further charge from *S2T*. However, the n-o *XCR* contact now closes, connecting point 12 to adjuster *A1P* and the positive terminal 17 of 13*C*. As electrons flow from 12*C* terminal 12, through the *XCR* contact and *A1P* to 13*C*, the potential at 12 rises as shown in Fig. 7*F*. If adjuster *A1P* (mounted on the same shaft with *B1P* in the tube-*Y* circuit) is turned clockwise to increase its resistance, point 12 rises more slowly.

As the d-c potential of point 12 rises, the a-c voltage wave of *S2T* (winding 4) is raised smoothly from position *L* in Fig. 7*G* to position *N* and then to positions *Q* and *S*.† As this *S2T* voltage

---

* In the following half cycle, 12*C* cannot lose its charge through 11 Rec, but 12*C* does discharge partially through *XCR* contact and 13*R*. This produces the jagged voltage wave across 12*C* shown in Fig. 7*F*. The resistance of 13*R* is less than that of 23*R* (in the tube-*Y* grid circuit) to make sure that tube *X* always fires first in each weld operation.

† Notice that the amount of *S2T* a-c voltage does not change; the phase position of this a-c wave does not change, for the positive crests at *P* and *U* remain directly above *M*.

106 ELECTRONIC MOTOR AND WELDER CONTROLS

wave rises, it fires tube X late in the half cycle as at O; a cycle or two later tube X is fired earlier, at R. Near the end of the weld, tube X is fired at S; here tubes X and Y are turned on completely. The number of cycles required for the tube-X grid potential to rise from position L to position S may be set by adjuster $A1P$ (and by $B1P$ in the tube-Y circuit), marked "Cycles to Final Heat."

When the welding current stops, relay XCR drops out; $12C$ again charges so as to hold point 12 negative, ready for the start of the next welding current.

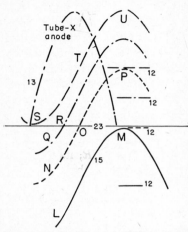

FIG. 7G. Position of $S2T$ voltage wave rises to fire tube X earlier.

7-3. Slope Control for Synchronous Combinations. When the slope control of Fig. 7E is used with all-tube circuits (such as the A151 or A152 combinations of Secs. 6–2 and 6–8), no relay contact is available to pick up relay XCR in the slope control. For use with such equipments, the slope control may be mounted on the door; a separate acces-

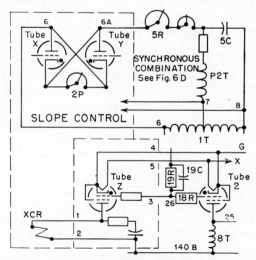

FIG. 7H. Tube Z is added when slope control is used with Fig. 6E.

sory, mounting thyratron tube Z, may be located inside the synchronous-timing section. Figure 7H shows this added tube-Z circuit, connected so that tube Z fires to pick up relay XCR at the same time that tube 2 fires in the synchronous timer to start the welding current. Here XCR is a d-c relay (like $1TD$ or $5TD$ in Figs. 2E and 2H and Figs. 6D and 6H). The slope-control circuits receive a-c power from terminals 6, 7 and 8 of transformer $1T$ in the synchronous timer's heat-control circuit. Operation within the slope control is the same as is described in Sec. 7-2.

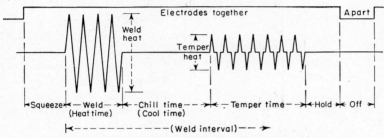

Fig. 7I. Changes of welding current during temper welding.

7-4. Temper Welding. While some metals (such as high-carbon steel) may be spot-welded by a single pulse of current, the resulting weld may be brittle and easily broken. To anneal (or temper) and strengthen such a weld, current again may be applied to the welding part before separating the welding tips. As outlined in Fig. 7I, such temper welding begins with an ordinary spot weld, usually of short weld time but large weld heat. Current then stops for a period called the chill time. When current flows again, this temper time may be much longer than the previous weld time; the temper heat (per cent current as set by the phase-shift heat-control circuit) usually is less than the weld heat. This temper heat causes the welded piece to cool more slowly, to "draw" the hardness from the weld and make it less brittle.

Such temper or post-heat welding may be obtained with a NEMA 5B pulsation-weld timer (Sec. 2-4), an ignitron contactor (Sec. 1-2) and a phase-shift heat control (Sec. 1-5), by adding suitable relays and extra time and heat adjusters, as is shown in Fig. 7J.* Here a temper-heat adjuster $A5R$ (a duplicate of $5R$) has been connected into the phase-shifting circuit; similarly a

* Such temper-weld accessories (CR7503–D149) have been provided for earlier types of synchronous timer.

temper-time adjuster $A21R$ (a duplicate of $21R$) has been connected into the heat-time circuit of the NEMA 5B timer. The n-c contacts of a relay CRS are used to short these added adjusters during the first flow of welding current. When this current stops, CRS operates so that its n-o contacts now short $5R$ and $21R$, but the n-c contacts are opened to let $A5R$ and $A21R$ control the temper heat and temper time.

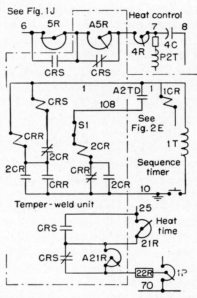

Fig. 7J. Temper-welding circuits added to sequence and heat controls.

Other relays are used in Fig. 7J to operate CRS at the end of the weld time. These relays operate from the 115 volts supplied by transformer $1T$ of the NEMA 5B timer; an unused contact of relay $A2TD$ (within the timer) is connected in series with the added relay $2CR$. Relay $A2TD$ closes its contact whenever welding current flows; if switch $S1$ is closed, to permit the temper weld, $A2TD$ picks up $2CR$ during the weld time. A $2CR$ contact picks up relay CRR, which seals itself closed through its CRR contact. At the end of the weld time, $A2TD$ drops out $2CR$; a n-c contact of $2CR$ now closes and, in series with the closed CRR contact, picks up CRS. Relay CRS now switches control of the heat from $5R$ to $A5R$ and switches control of the time from $21R$ to $A21R$. After the chill time, as set by the cool-time adjuster

SLOPE CONTROL, TEMPER AND FORGE 109

$31R$ (in Fig. $2E$) the NEMA 5B timer starts the second flow of current; this current is controlled by $A5R$ and $A21R$. The weld-interval adjuster $41R$ (in Fig. $2E$) is set so that $4TD$ is picked up during the temper time, to permit the hold time to begin when $A2TD$ drops out at the end of the temper time.

7–5. Temper Welding with Synchronous Combination (CR7503-A160). This control equipment is like the combination described in Sec. 6–9 except that its sequence timer (or top section) may provide a spot weld, a temper weld, or a pulsation weld (of not more than two pulses). Figure $7K$ shows this sequence-timer circuit; it may be used to replace tubes 21 to 25 in Fig. $6H$. Note that tube 29 is added, with relays $9TD$, $A9TD$ and $A4TD$; temper-heat and temper-time adjusters are added at the bottom of Fig. $7K$, switched by relay $A4TD$. Tube 23 of this sequence timer fires the synchronous timer as in Fig. $6H$; its action is described in Sec. 6–10 and should be reviewed here.

Before the starting switch closes, thyratrons 21, 23, 24, 25 and 29 have no anode voltage, so they cannot pass current. However, note that capacitors $11C$, $41C$ and $51C$ are fully charged, holding negative the grids of tubes 21, 24 and 25; the tube-23 and tube-29 grids also are held negative, through $1TD$ and $4TD$ contacts. Capacitor $31C$ is discharged, through contacts $6TD$ and $9TD$.

After the squeeze time, relay $1TD$ fires tube 23; the synchronous timer fires the ignitrons and starts the weld time. Since relay $A4TD$ is not yet picked up, only the resistance of $5R$ is in circuit with $4R$ (at lower left in Fig. $7K$) to shift the phase of $P2T$ to set the weld heat; $35C$ discharges only through $35R$ and $32P$, to control the weld time. During this weld time, tube 2 fires in the synchronous panel so that transformer $8T$ produces a grid voltage which turns on triodes $22A$ and $22B$. Electrons flow up through tube $22A$, charging $31C$ (since $6TD$ contact now is open) and turning off tube 23. This has no effect on the weld time; when once started by tube 23, the synchronous panel completes its weld time even though tube 23 has fired for only a half cycle.

When $1TD$ starts tube 23 and the weld time, another $1TD$ contact 45–70 (near tube $22B$ in Fig. $7K$) disconnects the tube-24 grid from the negative potential at 70. About one cycle later, the action of the synchronous panel makes tube $22B$ pass current, holding grid 45 negative during the entire weld time. (Since the time constant of $41C$ and $42R$ is about 3 cycles, tube 24 gets no chance to fire instantly when $1TD$ operates.) Since tube 24 is not firing during this weld time, relay $A4TD$ is not picked up, so its n-c contacts are shorting the temper-heat and temper-time dials on the synchronous panel.

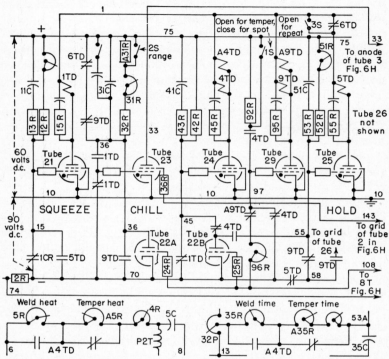

Fig. 7K. Temper-welding circuits in synchronous combination (CR7503-A160).

7-6. Chill and Temper Times. At the end of the weld time, the tube-$22B$ current stops (shown at P in Fig. 7L). Several cycles later tube 24 picks up relays $4TD$ and $A4TD$, at Q. If switch $1S$ is closed for a spot weld only, tube 29 fires at once when $4TD$ operates; the $9TD$ contact 36-70 prevents tube 23 from starting the synchronous panel a second time.

However, if $1S$ is open for a temper weld, then a $4TD$ contact (above tube 29) connects $92R$ in series with $96R$. As shown at R in Fig. 7L, these act as a voltage divider; $96R$ is adjusted so that point 97 (grid of tube 29) is kept about 12 volts more negative than cathode 10, preventing tube 29 from firing during the chill time. Meanwhile, relay $A4TD$ has operated; its contacts now

SLOPE CONTROL, TEMPER AND FORGE

short the weld-time and weld-heat dials, to let the temper-time and temper-heat dials control the impulse of current yet to come.

Meanwhile, at the end of the weld time the tube-22A current has stopped, so point 36 rises as 31C gradually discharges at a rate set by chill-time adjuster 31R. After this chill time, tube 23 fires again; the synchronous panel permits another impulse of current at the welder, but with new values of heat and time as set by dials A5R and A35R. Again tube 2 fires and transformer 8T turns on triodes 22A and 22B. Just above tube 22B note that

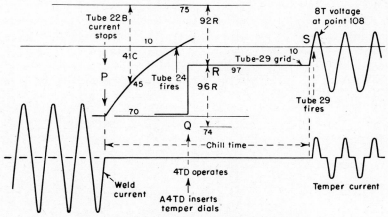

FIG. 7L. Temper-weld circuit action in Fig. 7K.

relay 4TD now has connected the tube-22B anode to the grid 55 of tube 25. When the temper current begins, the voltage of transformer 8T also overcomes the 12-volt bias at point 97 (at S in Fig. 7L) so as to fire tube 29. The A9TD contact opens above 96R, to prevent turning off tube 29. A 9TD contact (36–70) prevents tube 23 from firing. Another 9TD contact opens (55–58); however, grid 55 still is held negative by tube 22B so long as temper current flows.

At the end of the temper time, point 55 rises as 51C discharges at a rate set by hold-time adjuster 51R. After this hold time, a 5TD contact drops out 1CR to separate the welder electrodes. Another 5TD contact (58–70) opens, letting the tube-26 grid rise during the off time. If switch 3S is open for repeat welding, and the start switch is held closed, relay 6TD picks up after the off time; the 6TD contact drops out 5TD, whose contact again picks up 1CR to start another weld operation.

If 3S is closed for nonrepeat welding, 5TD remains picked up so long as the starting switch is held closed; the electrodes remain open. Relays 1TD, 4TD, $A4TD$, 9TD and $A9TD$ drop out. The starting switch must be opened to reset 5TD before another weld operation can begin.

7-7. Forge Welding. While large pressure may be needed at the electrode tips to force together heavy pieces being welded, this pressure may decrease the electrical resistance between the surfaces of these pieces, so that it becomes difficult to produce the heat needed for welding unless extremely large currents are used.*
To produce greater welding heat, some welding operations use low electrode pressure during most of the time of current flow; thus greater resistance remains at the welding surface. Just as the welding current stops, a second solenoid valve applies increased pressure to the electrodes, to force or forge the molten surfaces into closer contact.

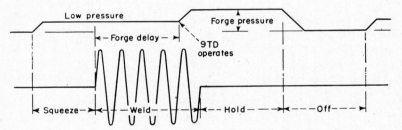

FIG. 7M. Changes of electrode pressure during forge welding.

Figure 7M shows such a forge-weld operation. To pick up the forge-pressure solenoid at the desired time (that may be anywhere during the weld or hold time), a separate time-delay circuit may be added with any sequence timer to provide the forge delay. Usually this forge-delay relay is made to begin timing at the end of the squeeze time.

7-8. Synchronous Combination with Forge Delay (CR7503-A164). To include the forge-delay feature in a complete control (such as that described in Sec. 6-3) the forge-delay circuit may be added into the sequence control shown in Fig. 6D. Figure 7N shows that this forge timer (tube 29, 91C, 91R, etc.) is a duplicate of those in Fig. 6D. Relay $A1TD$ is added, and is picked up by tube 21 at the end of the squeeze time; the $A1TD$ contact opens

* See footnote, Sec. 1-1.

to start the forge delay. When 91C has discharged (at a rate set by forge-delay adjuster 91R) so as to fire tube 29, relay 9TD picks up the solenoid valve to apply forge pressure. This forge-delay action occurs without affecting the weld-time action (of the

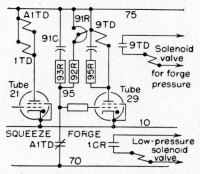

Fig. 7N. Forge-welding circuit in synchronous combination (CR7503–A164).

synchronous timer in Fig. 6E), or the hold time in Fig. 6D. At the end of the hold time, 1CR drops out the low-pressure solenoid and also turns off tube 21; at once $A1TD$ drops out the forge-pressure solenoid.

CHAPTER 8

G–E SEAM–WELDING CONTROLS

The welding controls previously described are used for spot welding—welding metals together at one spot, then separating the electrodes and bringing them together again before welding at another spot.

8–1. Seam Welding. In contrast, seam welding makes a continuous row of spot welds without raising the electrodes off of the work. The electrodes are usually a pair of copper-alloy rolls or wheels which turn as the work passes between them. Often the welds are made so close together that they overlap, producing a continuous weld that is airtight. This usually requires from 4 to 20 welds per inch, depending on the metals used.

Sometimes mild steel is welded by passing a steady flow of current between the wheel electrodes. However, most seam welds are made by letting the welding current flow in short pulses. For example, current may flow for 3 cycles, then stop for 2 cycles, and continue this "3 on–2 off" pattern as the pieces are stitch-welded together. For such a weld, the dials on the electronic control are set for 3 cycles heat time and 2 cycles cool time. In addition, a phase-shift dial may select the per cent heat, to control the amount of welding current during each cycle of heat time. Since the heat time and the cool time may be set as short as one cycle each, such precision timing is done by all-tube circuits.*

A sequence timer may not be required for a seam weld. The foot switch operates a solenoid valve to bring the wheel electrodes onto the work, then a pressure or limit switch closes to start the electronic seam-welding control. Welding current flows in pulses until the foot switch is released.

8–2. Seam-welding Control (CR7503–B120). Figure 8A shows a seam-welding combination. (The upper panel permits this timer to be used for spot or pulsation welding in addition to seam

---

* For earlier types of seam-welding controls see Chute, "Electronic Control . . .," Chap. 23.

welding; this complete combination is described in Sec. 8–7.) The middle panel contains the seam-welding-timer circuits next described. Many of these electronic circuits are the same as those used in the spot-welding combinations described in Chap. 6. The lower portion of wiring diagram Fig. 6E (including ignitrons

Fig. 8A. Synchronous seam-welding combination (CR7503–C121).

19 and 20, fired by thyratrons 17 and 18) is used also in the seam-welding control; the ignitron power panel is the lower section shown in Fig. 8A.

The diagram of the synchronous seam-welding section (center of Fig. 8A) is shown in Fig. 8B; for seam welding, this diagram replaces the upper portion of Fig. 6E, and is like it in many portions.

In brief, after the welder electrode wheels have pressed onto the work, the starting contact closes (at upper left in Fig. 8B);

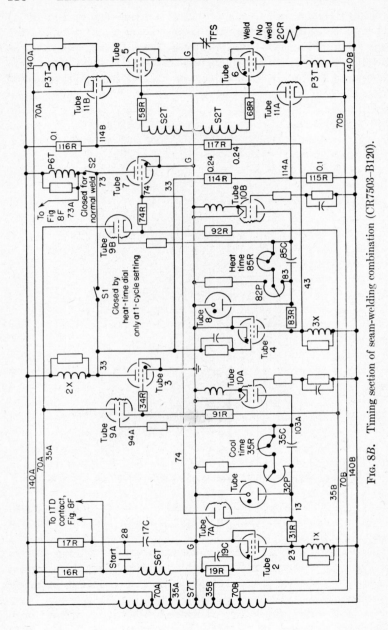

Fig. 8B. Timing section of seam-welding combination (CR7503–B120).

this fires tube 2, starting a cool time. After this cool time set by 35R, tubes 9A, 3 and 4 fire, instantly turning off tubes 11A and 11B (at the right); this lets tubes 5 and 6 fire, so that transformer 3T lets the main thyratrons fire the ignitrons (tubes 17, 18, 19 and 20, lower portion of Fig. 6E) to let welding current flow. After the heat time set by 85R, tubes 9B and 7 fire; above tube 7, transformer 6T turns off tube 2, resetting tubes 3 and 4. Tubes 11A and 11B pass current, turning off tubes 5 and 6 to stop the welding current. When tube 2 is turned off by $S6T$, tube 7A passes current and turns off tube 7 so that tube 2 again may fire to start another cool time. All these actions are repeated until the starting switch is opened.

8–3. Conditions before the Weld. As was mentioned in Sec. 6–3, the a-c welding power, supplied at the bottom of Fig. 6E, passes through a circuit breaker and the ignitrons to apply voltage to the welding transformer W. A control switch connects power to 1T and other transformers. After allowing 5 minutes for warming the tubes, time-delay relay TD closes the circuit to 7T, whose secondary $S7T$ supplies voltages at the left in Fig. 8B. Notice that $S7T$ is arranged like the anode transformer of a two-tube rectifier, for it has center tap G (grounded) and end taps 140A and 140B. Let us call half cycles $A1$, $A2$, $A3$, etc., those half cycles when point 140A is more positive than midtap G; these are the only half cycles when tubes 7A, 9A, 3, 9B, 7, 11A or 5 can fire. Similarly, call the other half cycles, $B1$, $B2$, $B3$, etc., when points 140B and 70B are more positive than midtap G; these are the only half cycles when tubes 2, 1, 10A, 4, 8, 10B, 11B or 6 can pass current. These half cycles are shown in Fig. 8C.

Before the starting contact closes in Fig. 8B, the tube-2 control grid (connected through 19R, $S6T$* and 16R) is at the potential of 70A, which is negative whenever the tube-2 anode is positive. Since there is no current through reactor 1X and tube 2, its anode 23 is at 140B potential. Most of the voltage from 140B to G appears across 31R and voltage-regulator tube 1; when 140B becomes positive enough (shown at D in Fig. 8C), tube 1 passes current so that point 13 is kept 105 volts more positive than G. During this half cycle $B1$, cool-time capacitor 35C is charged by

---

* Transformer winding $S6T$ furnishes no voltage at this time. The voltage across 19C (charged by tube-2 grid current in half cycle $A1$) drives the tube-2 grid negative before the start of half cycle $B1$.

electrons flowing from G through tube $10A$ into $35C$.* During the following half cycle (such as $A2$) when points $140B$ and 13 become negative, the charge on $35C$ drives the tube-$9A$ grid so negative that tube $9A$ passes no current through its cathode resistor $91R$. When $140A$ is positive, the tube-3 control grid is at the negative potential of $35B$ so tube 3 cannot fire. Since its anode 33 is now at $140A$ potential, the grid of tube 4 is thus kept negative during B half cycles so that tube 4 cannot fire.

Notice that the circuit including tube 4, $3X$, $83R$, tube 8, $85C$, tube $10B$ and tube $9B$ is exactly like the tube-2, tube-1, tube-$9A$ circuit described above. Therefore, so long as tube 4 is not firing, current may flow through tube $10B$ to charge heat-time capacitor $85C$ to the 105 volts regulated by tube 8; tube $9B$ passes no current through $92R$, so tube 7 cannot fire (with $S2$ closed for normal pulsing seam weld).

When tube 3 is not firing, this lets tube $11A$ pass current to prevent tube 5 from firing; let us describe this in detail. In Fig. $8B$, notice that the tube-3 anode 33 connects through $114R$ and $115R$ to $140B$. Resistor $114R$ has more ohms than $115R$; until tube 3 fires, their junction point $114A$ (grid of tube $11A$) is about 80 volts above the potential of point $140B$. However, the cathode of tube $11A$ is at $70B$ potential (70 volts above $140B$) so grid $114A$ is about 10 volts more positive than cathode $70B$ (during half cycle $A2$); electrons flow from $70B$ through tube $11A$, $58R$ and transformer $S2T$ to G. In this way the tube-$11A$ anode (which is also the grid of tube 5) is held closer to the negative potential of $70B$ so that tube 5 cannot fire. Although $S2T$ always produces a voltage that tries to fire tube 5, this $S2T$ voltage is wasted across $58R$ because of tube-$11A$ current.

Similarly, until 4 fires, its anode 43 is at $140B$ potential. Between 43 and $140A$, resistors $116R$ and $117R$ divide the voltage so that point $114B$ (grid of tube $11B$) is more positive than cathode $70A$ (during half cycle $B1$). Electrons flow from $70A$ through tube $11B$, $68R$ and $S2T$ to G; the tube-6 grid is held negative so that tube 6 cannot fire.

Until tubes 5 and 6 fire, transformer windings $P3T$ cannot let thyratrons 17 and 18 fire the ignitrons 19 and 20, shown in the lower part of Fig. $6E$. Tubes 5 and 6 in Fig. $8B$ act exactly like tubes 7 and 8 in Fig. $6E$, and as described in Sec. 6–7.

* Tube $10A$ (also $10B$) is grid-controlled like tube $5A$ mentioned in the footnote, p. 87.

In Fig. 8B we shall see that the ignitrons cannot fire, to let welding current flow, unless tubes 3 and 4 are firing. When tubes 3 and 4 do not fire, there is no welding current.

8-4. Timing the Seam Weld. Let us set the control dials in Fig. 8B to produce a seam weld made by 3-cycle pulses of welding current, separated by 2-cycle cool times. Heat-time dial 85R is set for 3 cycles, 35R is set for 2 cycles; switch $S1$ (upper center) is open, $S2$ is closed. The resulting current waves are shown in the lower part of Fig. 8C. Notice that a cool time occurs before the first flow of welding current.

With the wheel electrodes pressing onto the work, the starting switch closes at E (upper left in Fig. 8C) and remains closed during the entire seam weld. This connects the tube-2 control grid to 17C and 17R. Since the voltage across 17C drives the tube-2 grid negative soon after the beginning of half cycle $B1$ (as is shown at F) tube 2 cannot fire until half cycle $B2$ begins.

When tube 2 fires (during B half cycles), its anode 23 becomes only 15 volts (arc drop) above G; this prevents further recharging of 35C. The glow in tube 1 goes out, for such a tube passes no current when the voltage across it is so low. Since point 13 also is only 15 volts above G, the charge on 35C holds grid 94A negative, so that tube 9A passes no current and tube 3 does not fire. Now 35C starts to discharge through 35R during the cool time. If set in the 2-cycle position, the resistance in 35R will keep 35C charged for several cycles, as shown at H in Fig. 8C. At J the charge on 35C no longer holds the tube-9A grid negative. Tube 2, firing during half cycle $B3$, continues to fire and hold anode 23 positive during the early part of half cycle $A4$,* so as to let tube 9A pass current through 91R. This raises the grid of tube 3 close to G so that tube 3 fires in half cycle $A4$. Thereafter, so long as tube 2 fires in each B half cycle, tube 3 fires in each following A half cycle.

When tube 3 fires, its anode 33 remains 15 volts above G during all of half cycle $A4$ and also, due to inductance 2X, during the early part of half cycle $B4$ so as to fire tube 4. As tubes 3 and 4 fire, they turn off tubes 11A and 11B, to fire tubes 5 and 6 and the ignitrons as is described in Sec. 8-6.

When tube 4 fires, its action is like that of tube 2 described above. The tube-4 anode 43 becomes 15 volts above G; the

* This hangover action of tube 2 depends on inductance 1X, as is described in Sec. 4-5.

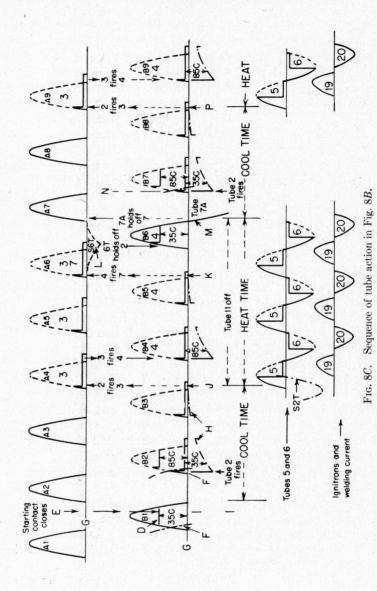

Fig. 8C. Sequence of tube action in Fig. 8B.

tube-8 glow stops. During the heat time, while $85C$ is discharging through $85R$, the charge on $85C$ holds negative the tube-$9B$ grid so that neither tube $9B$ nor tube 7 passes current. If heat-time adjuster $85R$ is set for 3 cycles, Fig. $8C$ shows that tube 3 is fired in three half cycles $A4$, $A5$ and $A6$ so that welding current flows for 3 cycles. But notice at K that $85C$ has discharged within 2 cycles; the extra cycle of heat time occurs while tube 7 is turning off tube 2, as is next described.

At K, $85C$ has discharged and tube 4 continues firing well into half cycle $A6$ so as to fire tube 7. Electrons flow from G through tube 7, switch $S2$ and $P6T$ to $140A$; $P6T$ is energized so that $S6T$ produces voltage in the grid circuit of tube 2. At L in Fig. $8C$, this $S6T$ voltage drives the tube-2 grid negative so that tube 2 cannot fire in half cycle $B6$, even though the starting contact still is closed.*

Since tube 2 does not fire during $B6$, its anode 23 returns to the potential of $140B$; capacitor $35C$ quickly recharges and tube 1 glows. Also, tube 3 cannot fire in $A7$, nor can tube 4 fire in $B7$; welding current stops after half cycle $B6$. At M in Fig. $8C$, the potential of points $140B$, 23 and 13 drops below G; since this drives the cathode of tube $7A$ (near tube 2 in Fig. $8B$) more negative than G, electrons flow from $140B$ or 13 through tube $7A$ to grid 74 of tube 7, through $74R$ and tube $9B$ to $35A$. In this way tube $7A$ passes current to pull down the potential of grid 74, to prevent tube 7 from firing during half cycle $A7$.† Since $S6T$ produces no voltage at N in Fig. $8C$, tube 2 is permitted to fire at the beginning of half cycle $B7$, to start the timing of another cool time. The action in half cycle $B7$ is like that in half cycle $B2$; tube 1, after glowing in half cycle $B6$, stops glowing in $B7$. Capacitor $35C$ begins to discharge; with $35R$ set for 2-cycle cool time, tube 3 does not fire in half cycles $A7$ or $A8$. However, at P the conditions

* Transformer $6T$ is of special design having an air gap in its core, so that $S6T$ produces voltage after $P6T$ no longer is energized. See footnote, Sec. 4–7.

If switch $S2$ is opened above tube 7, $P6T$ cannot be energized; tube 2 never is kept from firing so long as the starting contact remains closed. There is no heat-time–cool-time action; welding current flows without interruption until the starting contact is opened.

† Since tube 4 fires in half cycle $B6$ in Fig. $8C$, tube $9B$ still is passing current during the early part of half cycle $A7$. If tube $7A$ is not used, tube 7 fires during half cycle $A7$ for a second time, so that $S6T$ holds off tube 2 during $B7$ as well as during $B6$. Without tube $7A$, it is not possible to obtain less than 2 cycles of cool time.

are like those at J, and tube 3 fires (followed by tube 4) to start another heat-time flow of welding current. So long as the starting contact remains closed, the heat-time and cool-time periods will continue as shown in Fig. 8C.

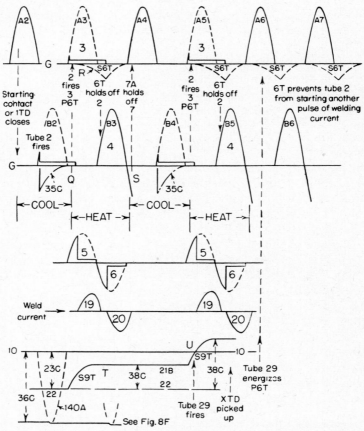

FIG. 8D. Tube sequence during one-cycle welds and during pulsation welds in Fig. 8F.

8–5. One-cycle Heat Time. At K in Fig. 8C where adjuster 85R is set for a 3-cycle heat time, it is seen that 85C discharges in less than 2 cycles. If 85R is set instead for 2-cycle heat time, 85C must discharge within 1 cycle, so that tube 3 fires in the half cycle immediately following tube 2; but tube 7 does not energize P6T until 1 cycle later.

To obtain yet faster action to produce a 1-cycle heat time, the heat-time adjuster $85R$ closes a separate contact (shown as $S1$, upper center of Fig. $8B$) to connect together the anodes of tubes 3 and 7. Through $S1$, tube 3 passes electrons to energize $P6T$. Figure $8D$ shows the timing action for a sequence of 1-cycle heat, 1-cycle cool. When the starting contact closes, tube 2 fires in half cycle $B2$ (similar to Fig. $8C$). Cool-time capacitor $35C$ discharges quickly so that, at Q, tube $9A$ passes current and fires tube 3. Tube 3 fires tube 4; tubes 5, 6 and the ignitrons are fired as is described in Sec. 8–6. However, with $S1$ closed, tube 3 also energizes $P6T$ in half cycle $A3$; $S6T$ produces a voltage to prevent tube 2 from firing in half cycle $B3$. Therefore tubes 3 and 4 are not fired during $A4$ and $B4$; this is the 1-cycle cool time. Capacitor $35C$ is again charged.

Since tube 4 has fired in half cycle $B3$ and $85C$ has discharged very quickly, tube $9B$ passes current at the beginning of half cycle $A4$, trying to fire tube 7. However, as in Sec. 8–4, tube $7A$ passes current at S so that the tube-7 grid is held negative. Tube 7 never fires while $85R$ is set for 1-cycle heat time. Since $P6T$ receives no voltage during $A4$, tube 2 may fire at the beginning of $B4$; $35C$ discharges so that tube 3 fires in half cycle $A5$, starting another heat time.

8–6. Tubes 11, 5 and 6. Phase-shift Heat Control. Whenever tubes 3 and 4 fire in Fig. $8B$, tubes 5 and 6 also fire, so that the ignitrons pass current to the welder (lower part of Fig. $6D$).

In detail, when tube 3 fires and its anode 33 remains 15 volts above G, during each A half cycle, this lowers the potential at all parts of the voltage divider $114R$ and $115R$; grid $114A$ is held so far below $70B$ that tube $11A$ passes no current. Tube 5 now is grid-controlled entirely by the voltage produced by $S2T$; this is a sine wave that holds the tube-5 grid negative during the early portion of each A half cycle. When the $S2T$ voltage swings positive, part way through each A half cycle, it fires tube 5 at this point. As is shown in the lower portion of Fig. $6D$, the primary winding $P2T$ is in the heat-control phase-shifting network made by $5C$, $4R$ and $5R$, connected in series across the 230 volts a.c. of $P1T$. By turning dial $5R$, the $P2T$ and $S2T$ voltages may be shifted in phase so as to control or change the point (within each half cycle of supply voltage) when tube 5 is to be fired in Fig. $8B$ (see Sec. 1–5).

When tube 5 fires, it reverses the voltage applied to $P3T$ above.

As is explained in Sec. 6–7, the $S3T$ voltage (in the tube-17 grid circuit in Fig. 6D) suddenly fires tube 17 so that ignitron 19 passes current to the welder.

During each B half cycle, when tube 4 fires and its anode 43 remains 15 volts above G, this lowers the potential at all parts of the voltage divider 116R and 117R; grid 114B is held so far below 70A that tube 11B passes no current. Tube 6 now is grid-controlled entirely by the $S2T$ voltage. In the same manner as the $S2T$ voltage fires tube 5 in the A half cycles, the opposite half wave of $S2T$ voltage fires tube 6 in the B half cycles. By turning 5R clockwise, tubes 5 and 6 are made to fire earlier within each half cycle, making tubes 17, 18, 19 and 20 fire earlier, thereby increasing the per cent heat at the weld.

8–7. Spot, Seam, Pulsation Welding Combination (CR7503-C121). This complete synchronous timer is shown in Fig. 8A; its control station appears in Fig. 8E. The center and bottom portions of the equipment are described in Secs. 8–2 to 8–6, and provide only the seam-welding feature. By adding the sequence panel shown as the upper portion in Fig. 8A, this equipment may be used for spot or pulsation welding,[2-3] as well as for seam welding. This added sequence circuit is shown in Fig. 8F, which includes portions of the seam-welding circuit at the right.

In Fig. 8F, tubes 21, 25, 26, 27 and 28 correspond to tubes 1, 5, 6, 7 and 8 in the NEMA 5B sequence timer of Fig. 2E, as described in Sec. 2–4. Tubes 27 and 28 furnish a regulated d-c voltage so that upper point 75 is at $+60$ volts above grounded point 10; bottom point 70 is at -90 volts, below 10. A selector switch $S4$ (having contacts above tube 29 and above the starting switch) is opened when a seam weld is desired, but is closed for a spot or pulsation weld. With $S4$ closed, a dial (21R, center of Fig. 8F) selects the number of spots or welding-current impulses that make a single pulsation weld. When 21R is set at the 1-spot position, a spot weld is obtained instead of a pulsation weld.

First let us close $S4$ in Fig. 8F and watch briefly as a spot weld is made. When the starting switch closes (lower right in Fig. 8F), relay 1CR picks up the solenoid valve to bring the welder electrodes together. After the squeeze time, tube 21 picks up relays 1TD and ACR. Above the starting switch, 1TD contacts close, to fire tube 2 in the seam-welding circuit at the right. As is outlined in Sec. 8–2 and Fig. 8B, tube 2 starts a cool time, after which the ignitrons pass current to the welder during the heat time.

(For a single spot weld, this heat time is called the weld time.) At the end of this heat time, tube 7 fires (upper right in Fig. 8F), applying voltage to $P6T$ to turn off tube 2 and end the flow of welding current. This voltage across $P6T$ is carried also into the sequence timer, at points 107 and 108, and is applied to $P9T$; $S9T$ forces electrons through tube $2B$ to charge capacitor $38C$.

Fig. 8E. Control station of spot, seam and pulsation panel (CR7503–C121)

If $21R$ is set for a single spot, $38C$ is charged fast enough to fire tube 29 at once and pick up relay XTD; a nearby XTD contact closes to let tube 29 keep $P6T$ energized, preventing tube 2 from starting a second pulse of welding current. At upper left, another XTD contact picks up relay QTD; a QTD contact opens below tube 25 to start the hold time. After the hold time, a $5TD$ contact drops out $1CR$, to separate the electrodes.

Now let us open $S4$ and watch a seam-welding operation. Above the starting switch in Fig. 8F, the $S4$ contact opens the circuit

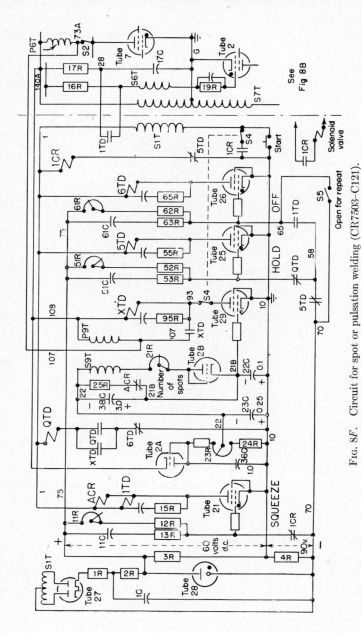

Fig. 8F. Circuit for spot or pulsation welding (CR7503–C121).

that would seal in $1CR$; the starting switch must be held closed during the entire seam weld. The squeeze time occurs as for a spot weld; then the $1TD$ contact fires tube 2 in the seam-welding timer at the right. Although tube 7 applies voltage to $P6T$ to end each pulse of welding current, and the voltage at 107–108 causes $38C$ to be charged, the $S4$ contact (opened above tube 29) prevents relay XTD from picking up QTD; the QTD contact does not start a hold time. Meanwhile the seam-welding panel continues to cause heat-time pulses of welding current, separated by cool times, as is described in Sec. 8–4. When the starting switch is opened, $1TD$ opens its contacts, ending the pulses of welding current; $1CR$ drops out, to separate the electrodes.

8–8. Number-of-spots Circuit. The central part of Fig. $8F$, with tubes $2A$, $2B$ and 29, controls the number of spots or pulses of welding current used to make a pulsation weld. If $21R$ is set in the one-spot position, a spot weld results.

Before closing the starting switch, but with $S4$ closed for a pulsation weld, notice that the grid of tube 29 is held negative by the constant d-c charge on capacitor $23C$; tracing from tube-29 cathode 10, across $23C$ to point 22, we pass $38C$ (shorted through $25R$ and ACR contact) to reach $21B$, grid of tube 29. Tube $2A$ acts as a rectifier to charge $36C$ and $23C$; during B half cycles (when $140A$, at upper right, is more negative than G), electrons flow from $140A$ through tube $2A$, $23R$* and $24R$ to 10, grounded to G. This $23C$ voltage is shown in the lower part of Fig. $8D$, and prevents tube 29 from firing. Figure $8D$ shows action during a pulsation weld made by two current pulses; $21R$ is set for two spots.

When the starting switch is closed in Fig. $8F$, relays $1TD$ and ACR operate after the squeeze time; the ACR contact opens below $38C$. At the right the $1TD$ contact closes, to fire tube 2 in half cycle $B2$ of Fig. $8D$. Tube 2 fires tube 3 to let welding current flow, as previously explained; tube 3 applies voltage to $P6T$ (since $S1$ is closed in Fig. $8B$, for 1-cycle heat time).[8–5] This voltage is carried also to points 107 and 108 to energize $P9T$ in Fig. $8F$; $S9T$ produces a half cycle of voltage that forces electrons from terminal 22 into $38C$; electrons return from $38C$ terminal $21B$ up through tube $2B$ and $21R$ to $S9T$. In the 2-spot position, most of the $21R$ resistance is shorted, so $S9T$ produces a large voltage

---

* Adjuster $23R$ calibrates the number-of-spots selector $21R$.

charge across $38C$ (shown at T in Fig. $8D$); terminal $21B$ is raised to a potential nearer to 10, but not close enough to fire tube 29. The charge remains on $38C$; meanwhile, $S6T$ has kept tube 2 from firing in half cycle $B3$, so tube 3 does not fire in $A4$ nor does it energize $P6T$. After the 1-cycle cool time, tube 3 again is fired, in half cycle $A5$; again voltage appears across $P6T$ and $P9T$; at U in Fig. $8D$, $S9T$ again forces electrons to charge $38C$. This time the $38C$ voltage becomes greater than the $23C$ voltage, so that terminal $21B$ (grid of tube 29) becomes more positive than 10 (cathode of tube 29). From G or 10, electrons pass through tube 29, $S4$ and XTD coil to 108 and the $S7T$ voltage at point $140A$. Relay XTD is picked up, and the XTD contact closes above tube 29. During half cycle $A6$ electrons flow from G or 10 through tube 29 and XTD contact, wire 107 and through $P6T$ to $140A$; by this tube-29 action, $S6T$ is made to produce voltage during all A half cycles after XTD has operated. Tube 2 cannot fire tube 3 to start another pulse of welding current; the pulsation weld is ended.

Another XTD contact closes (above tube $2A$ in Fig. $8F$) to pick up relay QTD, which seals itself in; a QTD contact opens below tube 25, to start the hold time. After the hold time, a $5TD$ contact drops out $1CR$ to separate the electrodes. If switch $S5$ is closed for nonrepeat welding, tube 26 cannot pick up $6TD$. (The $1CR$ contact has turned off tube 21 to reset ACR and $1TD$, so all tubes and relays have reset, except QTD and $5TD$.) Another weld operation cannot begin until the starting switch has been opened, to reset QTD.

If $S5$ is open, for a repeat weld, a $5TD$ contact (below tube 29) opens to start the off time; after the off time, tube 26 picks up $6TD$. The $6TD$ contact drops out QTD, resetting $5TD$. If the starting switch still is closed, $1CR$ again is picked up to start another welding operation.

CHAPTER 9

WESTINGHOUSE SYNCHRONOUS COMBINATIONS

The equipment next described will control a welding machine in the same manner as those in Chap. 6. The details of this circuit, as designed by another manufacturer, are of interest.

FIG. 9A. Synchronous spot-welding combination. (*Westinghouse Electric Corporation.*)

9-1. Westinghouse S2H Spot-welding Timer. This spot-welding control equipment includes a NEMA 7B sequence timer and a synchronous timer with heat control, firing an ignitron contactor. Figure 9A shows these parts grouped into a single case for mounting at the welding machine. The diagram of the sequence timer is Fig. 9C; this control fires the synchronous-timing and ignitron power circuits in Fig. 9D.

When the starting switch closes (at upper left in Fig. 9C), relay 5CR picks up the solenoid valve, to bring together the welding tips; other 5CR contacts seal around the starting switch and start the squeeze time. After the squeeze time, tube 21 picks up relay 1TD, whose contacts appear in Fig. 9D, and fire tube 11; the synchronous timer fires the ignitrons as is described later. After the weld time, tube 13 in the synchronous timer picks up relay 4CR of the sequence timer. The 4CR contacts open below tube 25 of Fig. 9C; after the hold time, tube 25 picks up 5TD, dropping out 5CR to separate the welding tips and to reset 1TD and the circuits of Fig. 9D. Another 5TD contact opens below tube 26. If switch 1S is set for nonrepeat welding, 5TD does not drop out if the starting switch still is closed. If 1S is set for repeat welding, tube 26 picks up 6TD after the off time; a 6TD contact resets 5TD. If the starting switch still is closed, 5CR again is picked up and the welding tips come together for another spot weld.

9–2. Westinghouse 7B Sequence Timer. The time-delay or grid circuit near tube 21 in Fig. 9C is like the circuit of tube 25 or tube 26; each of these timing circuits operates on the regulated voltage across tubes 19 and 20. All parts of the circuit shown in Fig. 9C connect to a-c voltage, received through transformer 12T; within each half cycle, however, voltage-regulator tube 19 or 20 serves to hold a constant voltage that acts like d.c. Thyratrons 21, 25 and 26 may fire during only those half cycles when 12T terminal 205 is more positive than center tap 277. During these A half cycles electrons flow through tube 19 and 77R so that point 285 remains 105 volts above 277, as is shown at M in Fig. 9B.* Similarly tube 20 passes current to hold constant the 206–277 voltage during the B half cycles as is shown at N.

Before the starting switch closes, electrons flow (during B half cycles) from cathode 277 to the control grid of tube 21, then

---

* During the following B half cycles electrons flow from 205 through 77R, 97R and metallic rectifier 11RX to 277; this action decreases the inverse voltage across tube 19 while its anode is negative. Similarly, during A half cycles, electrons flow from 203 through 78R, 98R and 12RX, to decrease the voltage across tube 20.

through $75R$, $74R$ and the $5CR$ contacts to bottom point 206; these electrons charge capacitor $31C$ so that its terminal is more negative at the tube-21 grid.* Since tube 20 is passing current during these B half cycles, $31C$ always charges to the same voltage, regulated by tube 20.

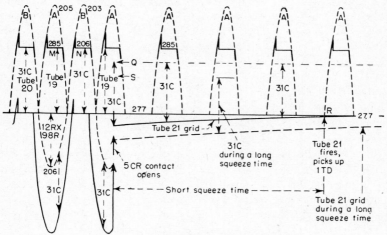

Fig. 9B. Squeeze-time action in Fig. 9C.

When the starting switch closes, picking up $5CR$ and the solenoid valve, the $5CR$ contact opens between $74R$ and point 206; $31C$ receives no further charge. (We are no longer interested in B half cycles, but must watch the action during A half cycles, for here the tube-21 anode is positive.) At once the 214 end of $31C$ rises to the potential of the slider of squeeze-time adjuster $11P$; this is shown at Q in Fig. 9B, when $11P$ is set for a short squeeze time. (The potential at Q remains constant during A half cycles, thanks to VR tube 19.) As $31C$ now discharges slowly through $75R$, the $31C$ voltage decreases; the tube-21 grid potential need not rise far before tube 21 fires at R. To obtain longer squeeze time $11P$ is turned clockwise; the $11P$ slider is at lower potential S so $31C$ must discharge for a longer time before the grid lets tube 21 fire. Tube 21 picks up $1TD$ to start the weld time as is described in Sec. 9–3.

During the weld time, tube-25 grid current flows (during B half cycles) through $105R$, $104R$ and the contacts of $4CR$ and $5TD$

* Since $74R$ has very little resistance compared with $75R$, $31C$ charges to nearly the entire voltage across tube 20.

to charge $37C$ to the regulated voltage across tube 20 (in Fig. $9C$). After the weld time, $4CR$ opens its contact; at once the 254 end of $37C$ rises to the potential of the $15P$ slider. After the hold time as set by $15P$, $37C$ has lost enough charge through $105R$ to let tube 25 pick up $5TD$. A $5TD$ contact drops out $5CR$; a $5CR$ contact turns off tube 21 so $1TD$ opens its contact (in Fig. $9D$,

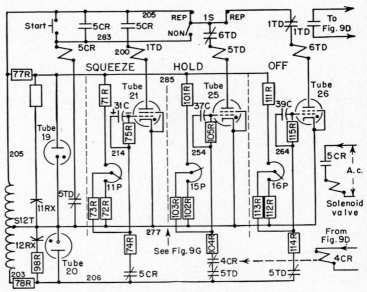

Fig. 9C. Spot-weld sequence timer for use with Fig. 9D.

top). This resets the synchronous-timing panel and drops out $4CR$; the $4CR$ contact in Fig. $9C$ now cannot affect tube 25 since a $5TD$ contact has opened below $4CR$. Another $5TD$ contact opens below tube 26 to let $39C$ discharge; if $1S$ is closed for repeat welding, tube 26 picks up $6TD$ after the off time as set by $16P$. A $6TD$ contact opens briefly above the $5TD$ coil; $5TD$ drops out and its contacts now hold negative the grids of tubes 25 and 26. A $5TD$ contact recloses near tube 19; if the starting switch still is closed, $5CR$ again is picked up, sealing itself in for another spot-weld operation.

9-3. Synchronous Spot Timer. As mentioned above, a $1TD$ contact closes to start the weld-time action in the upper part of Fig. $9D$. Main welding power enters at the bottom of Fig. $9D$ and must pass through ignitrons 5 and 6 to reach the welding transformer W. The ignitrons are fired by thyratrons 3 and 4 whenever tubes 1 and 2 are fired by the weld-time circuit next described. Control power is supplied through transformers $1T$, $3T$ and $7T$; several $S7T$ windings give voltages in the upper part of Fig. 9D.

Before $1TD$ operates to start the weld time, the control grid of tube 11 receives from $S7T$ an a-c voltage that is negative when

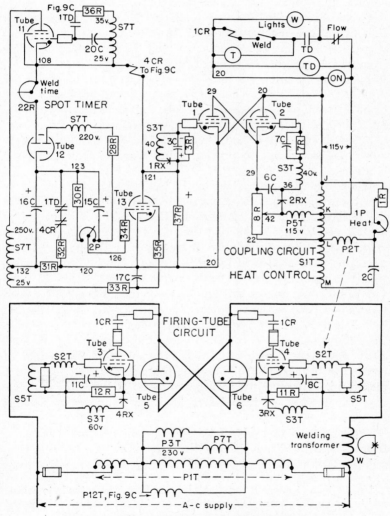

Fig. 9D. Circuit of Westinghouse synchronous timer.

the tube-11 anode is positive.* Tube 11 does not fire; its cathode 108 is at the potential of $S7T$ tap 132 so no voltage is applied to $4CR$ coil or across tube 13 (anode to cathode). However, another $S7T$ winding (near tube 12) forces electrons through $28R$, $2P$, $30R$ and tube 12 so as to charge $15C$; part of the voltage across $15C$ is selected by $2P$ to serve as a bias to hold negative the tube-13 control grid. Similarly, above tube 13 the $S3T$ voltage forces electrons through $3R$ and disk rectifier $1RX$ so as to charge $3C$; this $3C$ voltage is a negative bias to prevent tube 1 from firing. Therefore $S1T$ (at right center) cannot force electrons to flow from its top terminal 20 through tube 1 and $8R$ to center tap 22; with no voltage across $8R$, $6C$ receives no charge through rectifier $2RX$. The tube-2 grid is held negative by the out-of-phase voltage of $S3T$, which also charges $7C$ by tube-2 grid current. Terminal 42 of $P5T$ is at the potential of 22, so $P5T$ receives the voltage (115 volts) between K and L of $S1T$; the $S5T$ windings supply out-of-phase voltages to prevent tubes 3 and 4 from firing the ignitrons.

When the $1TD$ contact closes in the tube-11 grid circuit (at the top of Fig. 9D), this connects $36R$ and $20C$ across the upper half of the $S7T$ winding. As a result, the tube-11 grid now receives a wave of a-c voltage (leading the tube-11 anode voltage by 150 degrees) that is positive only at the start of the half cycle, so that tube 11 fires (at D in Fig. 9E) just as its anode becomes positive. At once the tube-11 cathode 108 rises to within 15 volts (arc drop) of its anode; this applies more than 200 volts (in positive half waves) between points 108 and $S7T$ tap 132. During each half wave when tube 11 fires, electrons flow from 132 through $31R$, $37R$ and the $4CR$ coil to 108 and through tube-11 anode to $S7T$. Since the resistance of $37R$ is large compared to that of the $4CR$ coil, most of the voltage appears across $37R$; $4CR$ is not picked up. However, this voltage across $37R$ is greater than the bias voltage of $3C$ so tube 1 fires, letting welding current flow, as described in Sec. 9–5.

During each half cycle of tube-11 current, electrons flow also from 132 into $16C$, returning through tube 12, the weld-time adjuster $22R$ and tube 11 to $S7T$.† Each pulse of tube-11 current

* When tube-11 anode is negative, electrons flow from cathode to control grid to charge $20C$; this voltage across $20C$ helps to drive negative the control grid before the tube-11 anode becomes positive (as at B in Fig. 9E).

† In Fig. 9D the weld-time adjuster is shown as a dial rheostat; instead, the Westinghouse panel uses a series of resistors that may be shorted separately

increases the voltage across 16C, thereby raising the potential of point 123. As 123 rises (one step each cycle as shown in Fig. 9E), the 15C voltage makes grid 126 of tube 13 rise an equal amount. If 22R is set for a short weld time (such as 3 cycles) the resistance in 22R is small; each pulse of tube-11 current greatly increases the charge in 16C. After the third pulse of tube-11 current, the 16C voltage has raised point 123 far enough that grid

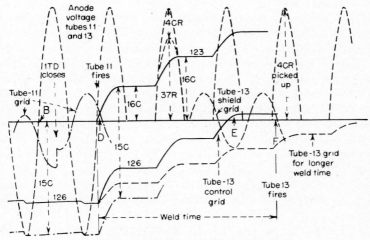

FIG. 9E. Weld-time action in Fig. 9D.

126 reaches tube-13 cathode potential (at E in Fig. 9E). However, its shield grid prevents tube 13 from being fired until the start of a half cycle,* so tube 13 does not fire until F. While the voltage across tube 13 and 37R decreases (to 15 volts arc drop), the voltage increases across the 4CR coil; 4CR is picked up, opening its contact (in Fig. 9C) to start the hold time. When tube 13 fires, the voltage across 37R becomes less than the bias voltage of 3C, so tube 1 does not fire; the welding current stops. After the hold time, when 1TD opens its contact and turns off tube 11, this drops out 4CR; contacts of 4CR and 1TD close (below tube 12 in

---

by toggle switches. Opening such switches increases the resistance and gives a longer weld-time adjustment; since 16C now receives less charge from each pulse of tube-11 current, more cycles of weld time will pass before the voltage across 16C rises enough to fire tube 13. This is shown in Fig. 9E by the broken line of tube-13 grid potential.

* Because of the a-c voltage phase-shifted by 17C and 33R, the tube-13 shield grid is driven negative soon after the tube-13 anode becomes positive.

Fig. 9D) so that 16C discharges quickly through 31R and 32R to be ready to time the next spot weld.

9–4. The Coupling Circuit. Through tubes 1 and 2 and nearby circuits in Fig. 9D, the timer circuit (described above) is coupled to the firing tubes 3 and 4. During the weld time, the voltage appears across 37R during only positive half cycles, to fire tube 1. Whenever tube 1 fires, tube 2 is fired in the following half cycle, as next explained.

When tube 1 fires, $S1T$ forces electrons from its terminal J through tube 1 and 8R to $S1T$ center tap 22. Most of this half cycle of $S1T$ voltage (J to L) appears across 8R. The upper half of this 8R voltage forces electrons into the 29 side of 6C; electrons return from 6C through rectifier $2RX$ to 42. So, while tube 1 fires, 6C receives a 150-volt charge that raises point 36. Enough of this 6C charge remains into the following half cycle so as to overcome the 40-volt $S3T$ winding and the 7C charge; tube 2 is fired.

When both half cycles of current thus pass through tubes 1 and 2 during the weld time, these tubes act like a switch closed between points 20 and 29. Since the $S1T$ voltage (J to L, 230 volts) is applied across 8R, the potential at 42 (halfway up 8R) is the same as at K (halfway between L and J), so no voltage remains across $P5T$ during the weld time.

9–5. Heat-control and Firing-tube Circuits. The grid circuit of firing tube 3 (in Fig. 9D) includes transformer windings $S2T$, $S3T$ and $S5T$; similar windings appear in the duplicate circuit of tube 4. During the weld time the $S5T$ windings produce no voltage (since $P5T$ receives no voltage, as mentioned above). However, $S2T$ and $S3T$ produce voltage every cycle. The 60-volt $S3T$ winding forces electrons through $12R$ and rectifier $4RX$ so as to charge 11C, which keeps a steady d-c voltage that holds negative the tube-3 grid; this bias voltage of 11C is shown in Fig. 9F. Transformer $2T$ is designed so that $S2T$ produces only a narrow peak of voltage each half cycle;* this peak is greater than the 11C bias voltage and drives positive the tube-3 grid. During the weld time, therefore, tube 3 fires ignitron 5 at that part of the a-c wave where the $S2T$ peak occurs; exactly one-half cycle later another

---

* This peaking transformer is explained in Chute, "Electronics in Industry," Chap. 28, and "Electronic Control . . .," Chap. 20. To make the voltage peaks occur in the desired portion of the a-c wave, the Westinghouse circuit connects $P2T$ in series with a tuning circuit (not shown in Fig. 9D) that includes a reactor and capacitor.

$S2T$ peak lets tube 4 fire ignitron 6. Figure $9F$ shows these actions during a 2-cycle weld time. Here the $S2T$ peaks occur near the middle of each half cycle of tube-3 anode voltage so that each ignitron passes current for less than a half cycle, to produce less than full heat at the weld. To increase the per cent heat or welding current, $1P$ is turned (at the right in Fig. $9D$) to

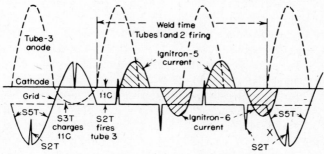

Fig. $9F$. Action in firing-tube circuit in Fig. $9D$.

decrease its resistance. This advances the phase of the a-c voltage applied to $P2T$* so that the $S2T$ peaks occur earlier in each half cycle, letting the ignitrons pass current during a larger part of each cycle.

At the end of the weld time, when tube 13 fires to remove the voltage across $37R$ (in Fig. $9D$), tubes 1 and 2 stop firing, to remove the voltage across $8R$. Transformer winding $P5T$ again receives 115 volts from $S1T$ (K to L), so the $S5T$ windings produce a voltage (shown at X in Fig. $9F$) that forces the grid negative when the anode is positive. Although the $S2T$ peak occurs as before, the $S5T$ voltage prevents $S2T$ from rising high enough to fire tubes 3 or 4. In this way, before and after the weld, $S5T$ prevents tubes 3 and 4 from firing the ignitrons. During the weld, the $S5T$ voltage disappears, letting the $S2T$ peaks fire tubes 3 and 4 and the ignitrons.

9-6. Westinghouse S3H Pulsation Welding Timer. This equipment is like that described above, except that a NEMA 9B sequence timer is substituted in the upper part of the enclosure, so as to produce either a spot weld or a pulsation weld. In addition to the squeeze, hold and off times discussed in Sec. 9-2, this pulsation sequence provides a cool time and a weld interval.†

* See Sec. 1-5.   † See Sec. 2-3.

The synchronous spot timer operates as before, but it is said to control the heat time—the length of each pulse of current during the pulsation weld.

Figure 9G shows the extra circuits to provide the cool time and weld interval; these may be inserted in the center of Fig. 9C. Notice that a 2TD contact now fires the synchronous timer of Fig. 9D, instead of the 1TD contact used previously.

If switches 2S are open (low in Fig. 9G) to give a spot weld, the 4TD contacts have no effect. The tube-22 grid already is at the positive potential of the 12P slider. When 1TD is picked up after the squeeze time, the 1TD contact closes below tube 22; at once this tube picks up 2TD whose contact fires tube 11 in Fig. 9D, letting the ignitrons fire during the heat time. After the heat time, relay 4CR opens its contact in Fig. 9G to start the hold time. Tube 22 does not drop out 2TD until after the hold time.

When switches 2S are closed, for pulsation welding, the 1TD contact makes tube 22 pick up 2TD after the squeeze time (since the 4CR n-o contact is open so that the tube-22 grid is positive). Another 1TD contact opens below tube 24 to start the weld-interval timer. When the synchronous timer (Fig. 9D) has completed the first heat time, relay 4CR closes its contact (below 2S in Fig. 9G); electrons flow from 277 through 1TD contact, through tube 22 (cathode to control grid), 95R, 93R, 2S, 4TD and 4CR contacts, to the regulated voltage at 206. This charges 35C and holds negative the tube-22 grid so that tube 22 drops out 2TD. At once the 2TD contact turns off tube 11 (in Fig. 9D), and 4CR drops out. The 4CR contact reopens below 2S; point 224 rises at once to the same potential as the slider of 12P. After the cool time as adjusted by 12P, 35C has discharged enough to let tube 22 again pick up 2TD; the 2TD contact again fires tube 11 in Fig. 9D to start a second heat time. So long as 4TD is not yet picked up, relay 2TD will continue to drop out during cool times and will fire tube 11 during heat times.

After the desired number of such heat-time pulses of welding current, as selected by weld-interval adjuster 14P, tube 24 picks up 4TD; one 4TD contact opens in the tube-22 grid circuit so that tube 22 will not again drop out 2TD. When 4CR is next picked up at the end of a heat time, it remains picked up and there will be no more heat times; the pulsation weld is finished. With the 4TD contacts already open, the opening 4CR contact starts the hold time. When 5TD is picked up after the hold time, 5CR

drops out (in Fig. 9C), separates the welding tips and resets $1TD$. The $1TD$ contact opens to drop out $2TD$ in Fig. 9G, to turn off tube 11 and drop out $4CR$. If $1S$ and the starting switch still are closed for repeat welding, the tips come together again after the off time, to start another pulsation weld.

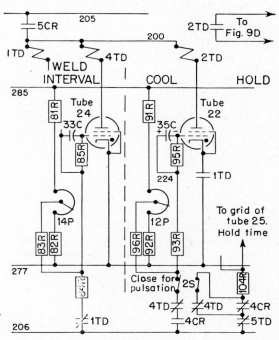

Fig. 9G. Pulsation-weld circuit, added in Fig. 9C.

9–7. Westinghouse Seam-welding Control. To provide a control equipment for a seam welder (as is described in Sec. 8–1) the spot-welding combination shown in Figs. 9A and 9D is adapted for this purpose by using a seam timer in place of the spot timer.* Figure 9H shows this seam-timer circuit, arranged to fire tube 1 of Fig. 9D; the coupling circuit, heat-control and firing-tube circuit remain the same as those described in Secs. 9–4 and 9–5.

Anode voltage for tubes 11 and 16 is received through $S8T$ at the left in Fig. 9H; this and five other $S8T$ windings furnish volt-

---

* For some seam welders no sequence timer is needed. The seam timer is mounted in place of the spot timer and is directly connected to the other panels below.

ages at all times. So long as relays $2CR$ and $3CR$ are picked up by the starting contact (shown at lower right and operated by the welding machine after the roller electrodes are together) the seam timer fires tube 1 during the heat times but turns off tube 1 during the cool times; whenever tube 1 fires, other circuits (in Fig. 9D) fire the ignitrons to let welding current flow between the roller electrodes.

Briefly, when the starting contact closes, a $2CR$ contact fires tube 16 at once, and the voltage across $48R$ fires tube 1; tube-16 current passes through tube 15 to charge $22C$. After the heat time, tube 11 is fired, turning off tube 1; tube-11 current passes through tube 12 to charge $17C$. At the end of the cool time, tube 14 fires only long enough to discharge $22C$; this fires tube 13 only long enough to discharge $17C$. Then tube 16 fires again to start the next heat time.

Notice that the grid circuit of tube 16 in Fig. 9H is like that of tube 11 in Fig. 9D, and described in Sec. 9–3. While the $2CR$ contact is open, the tube-16 grid is driven negative (while its anode 100 is positive) by part of the $S8T$ voltage and the charge on $23C$. However, when $2CR$ closes, this connects $46R$ and $23C$ across the other part of $S8T$; the tube-16 grid now receives a wave of a-c voltage that is positive only at the start of the half cycle. So, when $2CR$ closes, tube 16 fires at the beginning of each cycle thereafter, just as its anode 100 becomes positive.

Let us see other circuit conditions in Fig. 9H before the starting contact closes. Since tube 16 is not firing, its cathode 121 is at the potential of bottom point 134; there is no charge on $22C$, so terminal 173 of $21C$ also is at 134 potential. However, $21C$ has a voltage charge at all times, so $3P$ may adjust the negative bias on tube 11; $S8T$ (winding 4) forces electrons into the top of $21C$ and they return from 173 through tube 15 to $S8T$. Therefore all parts of $3P$ and the tube-11 grid are much more negative than 134. Tube 11 is not firing, and its cathode 120 is at 134 potential. This point 120 connects also to cathode 20 of tube 1, holding 20 at the potential of 134. Meanwhile tube 1 cannot fire, for its grid is more negative than 134, because of the charge on $3C$, produced by $S3T$ and rectifier $1RX$.

Since tube 11 is not firing, there is no charge on $17C$; terminal 123 of $16C$ also is at 134 potential. However $16C$ is charged at all times, so $2P$ may adjust the negative bias on tube 14; $S8T$ (winding 3) forces electrons into $16C$ that return from 123 through

WESTINGHOUSE SYNCHRONOUS COMBINATIONS 141

tube 12 to $S8T$. Therefore all parts of $2P$ and the tube-14 grid are much more negative than 134. These voltages are shown in Fig. $9I$. Similarly $19C$ is charged at all times (by $S8T$ winding 6 and rectifier $4RX$) to hold negative the grid of tube 13.

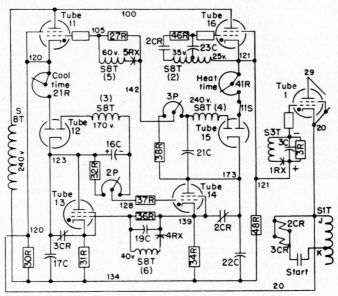

Fig. $9H$. Seam-welding timer for use with Fig. $9D$.

9–8. Heat and Cool Times. When the roller electrodes are together and the starting contact picks up $2CR$ and $3CR$ (at lower right in Fig. $9H$), a $2CR$ contact fires tube 16 as is mentioned above. At once the tube-16 cathode 121 rises to within 15 volts (arc drop) of anode 100. Large voltage appears across $48R$; point 121 now is so positive that it overcomes the $3C$ voltage and fires tube 1. At this same instant, tube 16 starts the heat time; electrons flow from 134 into $22C$, and from 173 through tube 15, $11S$,* $41R$ and tube 16 to 100. During each cycle of heat time these electrons increase the voltage across $22C$; the amount of this voltage rise depends on the setting of heat-time adjuster $41R$. (Rectifier tube 15 will not let $22C$ discharge through $48R$ and $41R$.) For the 2-cycle heat time shown in Fig. $9I$, each cycle of tube-16 current increases the $22C$ voltage by a large amount; after 2 cycles

* If $11S$ is opened, tube 1 fires steadily so long as the starting contact is closed. There is no interruption of the welding current, so a continuous seam results.

point 173 has risen to D, and the corresponding potential at $3P$ slider 142 could fire tube 11 at E, midway in a half cycle. To let tube 11 fire only at the start of a half cycle, a half wave of a-c voltage is added at the tube-11 grid; here $S8T$ (winding 5) forces electrons through $27R$ and rectifier $5RX$. Voltage appears across $27R$ only during half cycles when tube-11 anode is positive; in this way the tube-11 grid is held negative until the end of the half cycle, as at F. Tube 11 does not fire until G.

When tube 11 fires, its cathode 120 rises to within 15 volts of anode 100. Since 120 now is at the same potential as tube-16 cathode 121, the cathode of tube 1 has risen higher than the tube-1 grid, held negative by the $3C$ charge. Tube 1 stops firing and the welding current stops; the heat time is ended. At this same instant tube 11 starts the cool time; electrons flow from 134 into $17C$ and from 123 through tube 12, $21R$ and tube 11 to 100. During each cycle of cool time these electrons increase the voltage across $17C$; the amount of this $17C$ voltage rise depends on the setting of cool-time adjuster $21R$. For the 2-cycle cool time shown in Fig. $9I$, the first cycle of tube-11 current raises the potential of $2P$ slider (grid of tube 14) to J. Notice that the anode voltage across tube 14 is provided entirely by the charge on $22C$. At this same time the anode voltage across tube 13 is provided by $17C$.

The second cycle of tube-11 current charges $17C$ further, so that tube 14 fires at K. At once $22C$ discharges, forcing electrons through $34R$ and tube 14. For a brief instant this electron flow produces a voltage across $34R$ that raises point 139 high enough to overcome the $19C$ bias voltage; this fires tube 13, letting $17C$ discharge through $31R$ and tube 13. After these capacitors discharge, points 123, 139 and 173 all have returned to the potential of 134. The voltage of $21C$ drives negative the tube-11 grid, so tube 11 cannot fire when its anode next becomes positive at L; the cathodes 120 of tubes 11 and 1 both remain at 134 potential. However, tube 16 fires as before (since $2CR$ still is closed in its grid circuit) and its cathode 121 is held positive, again firing tube 1 to cause a second flow of welding current. The tube-16 current again charges $22C$ during another heat time. So long as the starting contact remains closed, tube 16 fires during every positive half cycle, but tube 11 fires only during the cool times. The difference in potential between tube-16 cathode 121 and tube-11 cathode 120 determines whether or not tube 1 shall fire, thereby causing welding current to flow.

WESTINGHOUSE SYNCHRONOUS COMBINATIONS 143

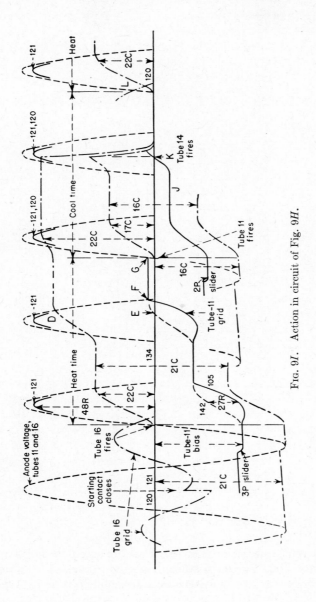

Fig. 9I. Action in circuit of Fig. 9H.

CHAPTER 10

WELDING WITH LIMITED POWER SUPPLY

Generally a resistance welder and its control operate from a single phase of the power system; such welders may draw large sudden loads that may disturb the power supplied to other users. In large plants this may not be a problem; in small plants remote from large power stations, large welders may not be used unless (1) the welding equipment includes series capacitors to decrease the load current, or (2) the welder is arranged to draw current from more than one phase, or (3) the welding equipment draws power gradually, storing energy that may be discharged suddenly to make the weld. These methods have been explained in an earlier text.* Recent methods of reducing welding-power demand by three-phase arrangements or by a frequency-changing system are described in Chap. 11.

10–1. Using Series Capacitors. Many large single-phase welders have been operating with series capacitors for many years.

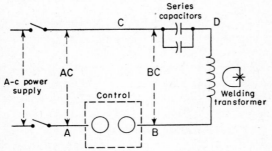
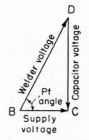

Fig. 10A. Simple connection of series capacitor in welder circuit.

Fig. 10B. Vector diagram of voltages in Fig. 10A.

The usual arrangement, shown in Fig. 10A, receives full line voltage at AC; line current flows through the capacitors, the welding transformer and the ignitrons, all connected in series. Although

---

* See Chute, "Electronic Control . . .," Chaps. 13, 25 and 26. See also the use of slope control, Sec. 7–1.

the same current flows through all three parts, the voltage across
the ignitrons is small (15 volts arc drop), while the voltage *BD*
applied to the welding transformer may be more than twice as
great as line voltage; the voltage *CD* across the capacitors is high
also. These voltages are shown in the vector triangle of Fig. 10*B*.
If the welder draws load current at 0.5 power factor, from a 440-
volt feeder, the welder voltage *BD* is twice *BC*, or about 850 volts.
For large-throat welders operating at 0.3 pf, the welder voltage
BD may be near 1500 volts, with 1400 volts across the capacitors
at *CD*. The welding transformer is built for this high voltage;
the control panel also must be designed for high-voltage operation.
A synchronous timer for use at high voltage, with or without series capacitors, is
described in Sec. 10–2.

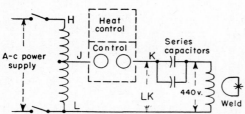
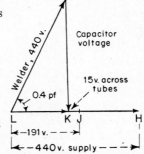

Fig. 10*C*. Autotransformer provides standard-voltage welder and capacitor circuit.

Fig. 10*D*. Vector diagram of voltages in Fig. 10*C*.

More recently series capacitors are used with standard 440-volt
welding transformers, ignitron contactors and heat controls,
merely by adding an autotransformer *HJL* as is shown in Fig. 10*C*.
Suppose that a 200 kva single-phase spot welder is to be operated
from a 440-volt feeder in a small plant, and that each spot weld
may require up to 400 kva at 0.4 power factor. To be able to
supply these load peaks, the power system may have to be
enlarged. Here, by adding the correct autotransformer and series
capacitors, the same weld may be made with only 160 kva drawn
from the present power system, and with the advantage of 1.0 pf
operation.

Figure 10*D* shows that this welder, operating at 0.4 pf, will
receive 440 volts at *LH*, if the autotransformer is designed to
supply 191 volts at *J*.* After allowing for 15 volts arc drop across

* If the 440-volt welder draws its load at 0.49 pf, it may operate directly
from a 230-volt power supply without an autotransformer.

the ignitrons, 176 volts remain at LK; since 176 volts = 0.4 × 440 volts, the welder receives 440 volts.

When using series capacitors with such a standard welder and ignitron contactor, a phase-shift heat-control accessory[1-5] is needed for varying the weld heat; taps on the autotransformer or welding transformer may not provide enough range. Within the limits of their proper use,* these low-voltage nonsynchronous controls work with series capacitors as satisfactorily as the high-voltage equipment previously needed.

10-2. High-voltage Spot-welding Control (CR7503–A148). This synchronous equipment may control a spot welder operating from a 2300-volt a-c power feeder, or it may be combined with series capacitors and a welder operating from a lower supply voltage. Figure 10E shows this control, having many high-voltage features in its lower section. The ignitrons† are fired by high-voltage thyratrons 7 and 10; these are cooled by air from a blower BL when the air in the cabinet is warm enough to trip thermostat T. Rectifier tubes 6 and 9 are added to produce d-c bias voltages for thyratrons 7 and 10. The fuses and anode resistors ($4F$, $5F$, $32R$, $36R$), relay $4CR$ and Thyrite resistor TH are selected for 2300-volt operation; door interlocks may be used to trip a line breaker to remove high voltage when the lower section is opened. The higher voltages appear only within this lower section or within the capacitor enclosure and at the primary terminals of the welding transformer, so these voltages cause no danger to the welder operator.

These high-voltage circuits are shown at the bottom of Fig. 10F. Main a-c power passes through the series capacitors, the welding transformer and ignitrons 8 and 11 when they are fired by thyratrons 7 and 10. The tube-7 grid circuit contains a hold-off bias across $29R$; $S8T$ produces a voltage to overcome this bias and fire tube 7. At all times $S6T$ forces electrons to flow through tube 6 and $29R$ so as to charge $15C$, holding point 87 about 200

* To provide enough voltage for firing the ignitrons, the autotransformer voltage JL must be more than 150 volts (except with very large welders). To permit use of a 440-volt welder in this manner, the load power factor must be above 0.3 pf; with shorted electrodes, the power factor may be only 0.07 lower than the power factor during the weld.

† At 2300 volts, rectifier-type ignitrons are selected; welding-type ignitrons are used at lower voltages. If operation with series capacitors applies more than 500 volts across the ignitrons, their ampere ratings must be decreased from those shown in Fig. 1L.

volts more negative than 80; this prevents tube 7 from firing until
$S8T$ produces voltage. Each time tube 2 fires, changing the voltage across $P8T$ (in the upper portion of Fig. 10F), $S8T$ produces
a voltage peak higher than 200 volts, so that it fires tube 7. Similarly, at that instant when tube 3 fires, changing the voltage across
$P9T$, $S9T$ produces a voltage peak that lets thyratron 10 fire
ignitron 11.

FIG. 10E. High-voltage synchronous control (CR7503–A148).

Control power enters the center part of Fig. 10F to energize the
small transformers and warm the tubes. After 5 minutes the time
relay TR applies voltage to $P2T$; the voltage of $S2T$ (at upper
left) now is rectified by tube 1 to produce 400 volts d.c. (filtered
by $1X$ and $1C$). Resistors $1R$, $1P$, $2R$ and $3R$ act as a voltage
divider to provide d.c. to the circuits of tubes 2, 3, 4 and 5. At
once tube 2 fires (as explained later), so that relay $3CR$ is picked

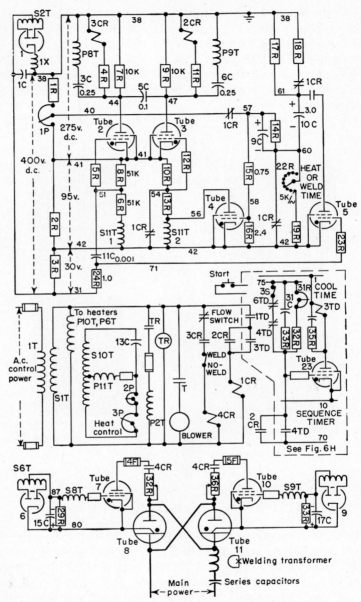

Fig. 10F. Circuit of high-voltage control (CR7503–A148).

WELDING WITH LIMITED POWER SUPPLY

up. If the water-flow switch and the weld–no-weld switch are closed (center of Fig. 10F), relay 4CR closes its contacts above the ignitrons so that welding current may flow after the starting switch is closed.

In Fig. 10E the upper panel is the sequence timer used as part of this spot-welder control. Its circuit is like that shown in Fig. 6H, except for the tube-23 circuit. Therefore Fig. 10F shows the tube-23 circuit, with its 3TD relay, for controlling this high-voltage spot welder.

When the starting switch closes, the 2CR and 4TD contacts are not closed (above tube 10 in Fig. 10F) so 31C has no charge; relay 3TD is picked up at once while the welding tips are coming together. After the squeeze time, the 1TD and 3TD contacts both are closed, so relay 1CR is picked up, operating its five contacts in the upper part of Fig. 10F. Section 10–5 explains how this lets tube 5 fire to start the heat time; tubes 2 and 3 fire alternately so that transformer windings $S8T$ and $S9T$ fire tubes 7 and 10 and the ignitrons permit welding current to flow, as explained above.

Tube 3 also picks up relay 2CR whose contacts close near the sequence-timer circuit; one 2CR contact connects the tube-23 grid to the negative potential at 70, turning off tube 23 so that relay 3TD opens its contact. Another 2CR contact closes around this 3TD contact to keep 1CR energized. After the heat time, tube 4 fires and stops the tube-3 current, dropping out 2CR; the 2CR contacts open, dropping out 1CR and starting the cool time. After 31C has discharged during the cool time as adjusted by 31R, tube 23 again picks up 3TD; the 3TD contact picks up 1CR to start a second heat time. In this manner there may be many heat and cool times during the pulsation weld, until weld-interval tube 24 picks up 4TD in Fig. 6H; when the 4TD contact closes, tube 23 cannot pick up 3TD to start another heat time. After the hold time, the welding tips separate.

If the sequence timer is set for a single spot weld, the 4TD contact closes immediately after the squeeze time, so that tube 23 cannot pick up 3TD a second time; 1CR operates only once.

10–3. Conditions before the Weld. In the upper portion of Fig. 10F notice that thyratron tubes 2 and 3 with their load resistors 7R and 9R are connected across 275 volts d.c. Since these are thyratrons, their control grids may prevent them from firing, but these grids cannot regain control so long as anode current flows. The tube-2 grid is held negative by the potential at point 51; because of the voltage divider $6R$ and $8R$, point 51 is nearly 50 volts below the point-41 cathode. Once during each cycle the transformer winding $S11T$ produces a voltage peak that drives point 51 more positive than 41; this fires tube 2 so that electrons flow from 31 up through $3R$, $2R$, tube 2 and $7R$ to 38. Some of these tube-2 electrons flow through $4R$ and the $3CR$ coil, picking up relay $3CR$. The tube-2 anode 44 remains 15 volts above cathode 41; thus 260 volts d.c. appear across $P8T$ and $3C$ in series. Capacitor $3C$ quickly is charged by this d-c voltage so that no voltage remains across $P8T$; although tube 2 is passing current steadily, $P8T$ receives no voltage.

The tube-3 grid is connected to a similar point 54 on voltage divider $10R$ and $13R$. However, until relay $1CR$ is operated to make a weld, one $1CR$ contact connects 54 to 42, holding the tube-3 grid 95 volts negative. Tube 3 cannot fire; its anode 47 remains at the top potential of 38, so no voltage is applied to $2CR$ or $P9T$.

Above tube 4, a $1CR$ contact connects $15R$ and $16R$ across the voltage (approximately 120 volts d.c.) between 42 and the slider of $1P$; the voltage across $16R$ holds the tube-4 grid positive so that tube 4 may fire and short-circuit the voltage peak produced by $S11T$ (winding 2). Other $1CR$ contacts connect point 42 to capacitors $9C$ and $10C$, and connect $10C$ through $18R$ to top point 38. (Another $1CR$ contact is open, preventing tube 5 from firing.) So $10C$ is charged to the 370 volts d.c. between points 42 and 38; $9C$ is charged to the 120 volts d.c. between 42 and the $1P$ slider. These voltages are shown in Fig. $10H$, explained in Sec. 10–5.

10–4. Tube Action during the Weld. After the squeeze time, relay $1CR$ is picked up. One $1CR$ contact opens to let tube 3 fire, as is next described, thereby starting the welding current. (The action of the other $1CR$ contacts and tubes 4 and 5 is described in Sec. 10–5.) We shall see that the movement of the $1CR$ contacts does not fire any tube; each tube is fired by a $S11T$ voltage peak, after the $1CR$ contact movement has made this possible.

At the left in Fig. $10G$ we see voltages in the tube-2 and tube-3 circuits (upper part of Fig. $10F$), before starting the weld. Since tube 2 is firing steadily, its anode 44 is close to its cathode 41; the voltage 38–44 has picked up relay $3CR$ and has charged capacitor $5C$. The voltage peak of $S11T$ (winding 1) drives the tube-2 grid positive once during each cycle of the a-c power voltage, but this has no effect since tube 2 has been fired by an earlier peak.

At A in Fig. 10G the 1CR contact opens between points 42 and 54; at once 54 rises until the tube-3 grid remains 50 volts below cathode 41. (Tube 4 cannot fire; its grid has been driven negative as described in Sec. 10–5.) When $S11T$* (winding 2) produces its next positive voltage peak, it raises grid 54 above cathode 41, firing tube 3. Electrons flow from 41 through tube 3 and 9R (also through 11R and 2CR) to 38. The potential at tube-3 anode 47 drops to 15 volts above 41. Since capacitor 5C cannot discharge at once, the voltage across 5C forces the tube-2 anode 44 far negative for a very short time (at C), stopping the tube-2 current.† This lets the tube-2 grid (now 50 volts negative) regain control to prevent tube 2 from refiring. Electrons then flow through tube 3 into 5C and through 7R to 38; quickly 5C loses its previous charge and lets the tube-2 anode 44 rise to the positive potential at 38; 5C becomes charged $(+)$ at 38, $(-)$ at 47.

Because tube 3 fires, lowering the potential at its anode 47, the voltage 38–47 suddenly is applied across 6C and $P9T$. While 6C is charging quickly to this new voltage, electrons flow from 6C through $P9T$ to 38, applying voltage to $P9T$ as shown at B in Fig. 10G. This makes $S9T$ produce a voltage peak (at E) that fires tube 10;‡ as explained earlier,[10–2] this in turn fires ignitron 11 to start the flow of welding current in the first, or leading, half cycle.

One half cycle later (at F), $S11T$–1 again drives point 51 above 41, firing tube 2. The tube-2 anode 44 returns to 15 volts above 41; the charge on 5C drives the tube-3 anode 47 negative at G, stopping the tube-3 current. This lets the tube-3 grid (now 50 volts negative) regain control to prevent tube 3 from refiring. Electrons flow through tube 2 into 5C and through 9R to 38; as

---

* When $S11T$–2 drives point 54 positive, $S11T$–1 drives 51 negative. Since the positive peaks fire tubes 2 or 3, only these positive peaks are shown, a half cycle, or 180 deg, apart. The position of these peaks (compared to the sine wave of supply voltage) may be adjusted by the per cent-heat dial, as explained later.

† This action by tube 3 to turn off tube 2 is similar to the inverter action described in Sec. 1–9 and Fig. 1N.

‡ Since the tube-2 current is stopped at this same time, the voltage across $P8T$ decreases as 3C discharges through 4R, 3CR coil and $P8T$; the resulting $S8T$ voltage drives the tube-7 grid further negative and cannot fire tube 7.

5C again reverses its charge, the tube-3 anode 47 returns to the positive potential at 38.

As tube 2 refires, the voltage 38–44 suddenly is applied across $3C$ and $P8T$; $3C$ increases its charge so electrons flow from $3C$ through $P8T$ to 38. This makes $S8T$ produce a voltage peak (at H) that fires tube 7; this in turn fires ignitron 8 to permit welding current to flow during the second, or trailing, half cycle.

Again one half cycle later (at J) a $S11T$–2 voltage peak fires tube 3; this repeated action turns off tube 2 and starts a second cycle of welding current. Figure 10G shows a 2-cycle heat time or weld time. As is explained in Sec. 10–5, the tube-4 grid becomes positive (at K), so that tube 4 may fire (at M) to short-circuit the next, or third, voltage peak of $S11T$–2. However, this does not prevent $S11T$–1 from firing tube 2 (at L), to fire the trailing ignitron 8 a second time. During the heat, or weld, time, we see that the $S11T$ voltage peaks continue to fire tubes 3 and 2 in turn, one half cycle apart, until tube 4 prevents further firing of tube 3. In any weld, tube 3 always starts the action, through $9T$, to fire ignitron 11; tube 2 always ends the action, through $8T$, firing ignitron 8.

Although the tube-2 current stops for half-cycle periods during the weld time, relay $3CR$ does not drop out; the current supplied by $3C$, discharging through $3CR$ coil and $P8T$, keeps $3CR$ energized. Similarly, relay $2CR$ is picked up when tube 3 first fires; when the tube-3 current stops for half-cycle periods, $6C$ discharges to keep $2CR$ energized. When tube 3 is prevented from firing in turn, $2CR$ then drops out.

As is shown at E in Fig. 10G, the ignitrons are fired only at those instants when the $S11T$ voltage peaks occur. To increase the per cent heat at the weld (by the phase-shifting method),[1–4] each ignitron is fired earlier within its half cycle of a-c anode voltage; this is done by advancing the position of the $S11T$ voltage peaks. Near the center of Fig. 10F, $P11T$ is shown in the phase-shifting network of $13C$, $2P$ and $3P$, receiving a-c voltage from $S10T$. To make the $S11T$ peaks occur earlier and to increase the weld heat, $3P$ is turned clockwise, decreasing its resistance. When the $S11T$ peaks are thus advanced, there is no change in the wave shapes shown in Fig. 10G except that the earlier firing of ignitrons 11 and 8 may produce a full sine wave of welding current.

10–5. The Heat-time Circuit. Whenever the voltage peak of $S11T$–1 forces point 51 positive, this change at 51 is faster than

WELDING WITH LIMITED POWER SUPPLY 153

capacitor 11C can charge through 24R; as 51 rises, the voltage across 11C makes point 71 rise higher than cathode 42 of tube 5. After relay 1CR is picked up in Fig. 10F and closes the anode circuit of tube 5, the tube-5 grid remains at the negative potential of point 31 until the next positive peak of S11T-1; through 11C,

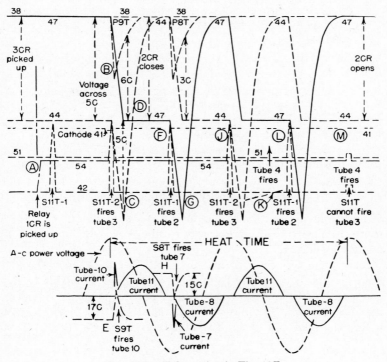

Fig. 10G. Tube action in Fig. 10F.

this peak fires tube 5, as is shown in Fig. 10H. At once point 61 is brought down to 15 volts above cathode 42; the 370-volt charge on 10C drives point 60 far below 42. Point 57, now disconnected from 40 by a 1CR contact, is held by 9C always 120 volts above 60 (since 9C and 14R have a very long time constant); however, 57 and grid 58 of tube 4 are driven negative (at P) so that tube 4 cannot fire. The next S11T-2 peak may fire tube 3, as described in Sec. 10-4.

When tube 5 fires, the heat-time capacitor 10C begins to discharge, forcing electrons to flow from 60 through 22R, 19R, tube 5 and the 1CR contact to 61. The amount of resistance in heat-time adjuster 22R sets the rate of discharge of 10C. In Fig. 10H, 22R is set for a 2-cycle heat time. As the 10C terminal 60 rises (at Q), the constant 9C voltage makes point 57 rise at the same rate; the tube-4 grid 58 rises also. After 2 cycles of time, points

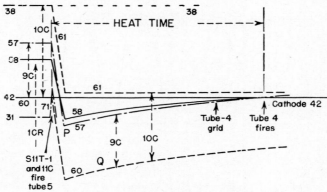

FIG. 10H. Heat-time action in Fig. 10F.

57 and 58 have risen close to cathode 42 so as to fire tube 4; tube 4 short-circuits S11T-2 and prevents tube 3 from firing again, as mentioned earlier.

Since tube 5 is fired because of a S11T-1 peak but tube 3 is fired one half cycle later by a S11T-2 peak, 10C begins to discharge a half cycle before the welding current begins. However, the discharge of 10C may fire tube 4 to end the timing action before the final half cycle of welding current flows.

CHAPTER 11

THREE-PHASE WELDING CONTROLS

Spot welders may demand a large current from a single phase of the a-c supply system; in some locations this may cause too much dip in supply voltage. To decrease the amount of this sudden change of voltage, various welding systems and machines have been introduced.* Most of these systems aim to distribute the single welding-transformer load among all three phases of the a-c power supply; this step alone may reduce by one-third the current in each line. Some welding machines include rectifiers whereby direct current may store energy in capacitors or in a special welding transformer;† the line current is drawn more steadily and in smaller amounts.

11-1. Multitransformer Welding Machines. Three-phase welding may refer to the use of three or more welding transformers in one welding machine; each transformer is a standard single-phase unit and supplies low voltage to its separate set of welding electrodes. One or more of these transformers are connected to each of the three phases so that the total welding load draws about the same amount of current from each of the three power lines.

As is shown in Fig. 11A, three standard two-tube ignitron contactors may control all of these welding transformers.‡ A three-pole relay (operated by a NEMA 3B timer[2-1]) may fire all ignitrons at once.

In some machines having many welding transformers, all electrodes may press onto the work together, but only half or one-third of the electrode tips pass current at one time. Usually all transformers are connected to the same phase but are energized in

---

* The use of series capacitors to decrease the load current of single-phase a-c welders is described in Chap. 10.
† Chute, "Electronic Control . . .," Chap. 26.
‡ The electrodes supplied from one phase should not touch the metal work close to electrodes supplied from another phase. While two ignitron contactors alone may control all three phases, three ignitron contactors must be used if they are phase-shifted to control the heat.

several groups so as to reduce the peak load. Here two or three ignitron contactors are used, fired in succession by a sequence timer that provides squeeze time, weld time 1, weld time 2, weld time 3, hold time and off time.

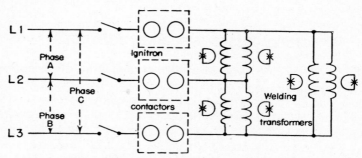

Fig. 11A. Single-phase welders arranged as a three-phase load.

11-2. Lower Frequency for Welding. To obtain further decrease in the line current required for making a weld, recent electronic controls change the 60-cycle power supply into a lower frequency such as 12 cycles per second; this 12-cycle current is applied to a welding transformer designed for this lower frequency. Let us see why lower frequency lets us weld with less line current.

We know well that, if a certain weld requires 150 kva or 350 amperes at 60 cycles (as in a Detroit automobile plant), that same weld can be made with about 90 kva or 200 amp, by a similar welder operating from a 25-cycle power supply (as in a Canadian-branch automobile plant near Detroit). In each case the weld requires 60 kw of power. To deliver this power to produce heat at the weld, the current must pass through the welding transformer and the high-current arms of the welding machine. At 60 cycles, this circuit resists the flow of current much more than it does at 25 cycles.* So the 60-cycle power feeder must furnish 350 amperes, to be able to produce 60 kw at the weld; from a 25-cycle feeder this same 60 kw can be produced with only 200

---

* In most welders, the current may be limited by the reactance of the secondary circuit more than by the resistance of that circuit and the resistance of the material welded. Since the reactance decreases as the power frequency decreases, a larger portion of the total kva is available (at lower frequency) to produce weld heat. The power factor of the welding machine is higher at 25 cycles than at 60 cycles.

amperes. If a 12-cycle power feeder was available, this weld would need less than 200 amperes. Still less current might be needed if the weld is made with direct current (or zero frequency), as in a metallic-rectifier system having no transformer in its low-voltage circuit.

To picture how the 60-cycle feeder voltage may be changed into 12-cycle power to make a weld, Fig. 11B shows (a) the 60-cycle voltage wave, and (b) the wave of low-frequency current at the weld. While this welding-current wave differs from a sine wave

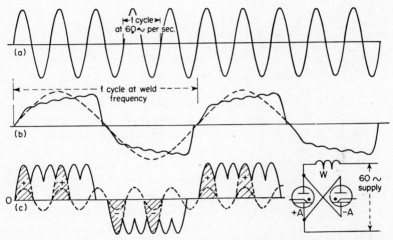

Fig. 11B. Using tubes to change 60-cycle power into 12-cycle power for welding.

(dashed line), it meets most welding needs. Here each current wave or cycle is as long as five cycles of the 60-cycle supply; there will be 12 such waves in each 60 cycles or one second, causing a welding-current frequency of 12 cycles per second. To produce this 12-cycle current wave, Fig. 11B shows (at c) how two tubes may be connected back to back between the 60-cycle feeder and the welding transformer W. If ignitron tube $+A$ is made to fire during cycles marked $(+)$, and tube $-A$ is fired during cycles marked $(-)$, notice that these shaded half cycles are above and below the 0 line at the right times to cause one alternation each five cycles (of 60-cycle supply). To produce the whole current wave shape of (b), we must add other pairs of ignitrons; these, connected in other phases of the 60-cycle supply, must fire so as to give the nearly-square voltage waves shown in (c). The welder current wave may look like this square voltage wave, or it may have the more gradual rise shown in (b) of Fig. 11B, and in Fig. 11F; this rise depends on the design of the welding machine.*

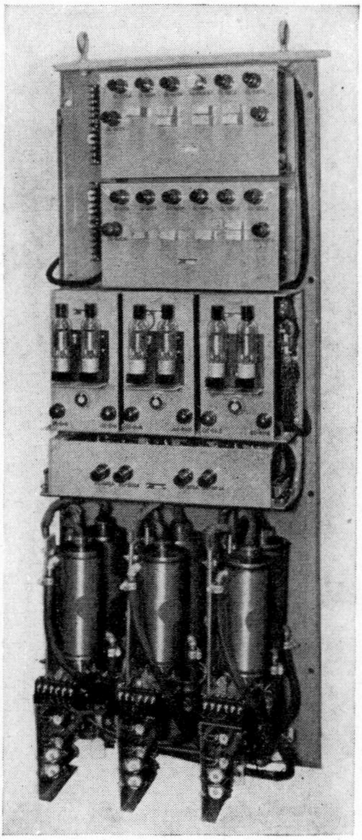

The welding transformer used in such a low-frequency system is built with enough core iron so that its magnetism (magnetic flux) does not build up to full value in less than $\frac{1}{25}$ sec. Therefore its flux is changing throughout the low-frequency half cycle, thereby producing secondary voltage during the entire half cycle.

11-3. Frequency-changer Control (CR7503–M101). Figure 11C shows an all-electronic control equipment that changes a three-phase 60-cycle supply into 12-cycle† power for a spot welder. A composite diagram appears in Fig. 11G, showing the general circuit arrangement, with references to the detailed diagrams of various circuit parts. Later we shall study these various timing, coordinating and firing circuits, including the smaller tubes shown in the upper part of Fig. 11C.

FIG. 11C. A three-phase frequency-changer welding control (CR7503–M101).

In the lower part of this equipment, notice the six ignitron tubes that carry the current to the welding transformer. These ignitrons are sketched in Fig. 11D; each of the three pairs is connected back to back in series with one primary winding of the welding transformer. While this transformer has but a single low-voltage secondary winding (connected to the welding tips),

---

* This gradual current rise often gives better welding results, less electrode maintenance, etc. Similar improvements may be obtained by slope control, described in Chap. 7.

† By internal switch adjustments, lower frequencies are obtainable, such as $6\frac{2}{3}$ or $8\frac{4}{7}$ cycles per second.

it has three primary windings wound on a single iron core. Figure 11E shows these three windings and the six ignitrons in their electrical three-phase arrangement in a triangle. We see that ignitrons $+A$ and $-A$ operate in phase A (between lines 1 and 2); ignitrons $+B$ and $-B$ operate in phase B (between lines 2 and 3).

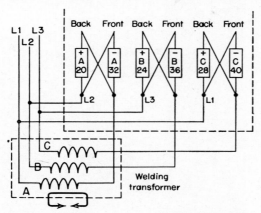

Fig. 11D. Connections to ignitron tubes of Fig. 11C.

We see that the voltage in phase B lags 120 deg behind the phase-A voltage; similarly, the phase-C voltage lags another 120 deg behind phase B. The waves in Fig. 11E show how tubes $+A$, $+B$ and $+C$ fire in turn, to produce voltage above the 0 line; tubes $-A$, $-B$ and $-C$ fire in turn to produce voltage below the 0 line. At any one time, only the $(+)$ tubes fire, or only the $(-)$ tubes; a $(-)$ tube must not fire while a $(+)$ tube is firing.

Since each of the three separate phase currents (controlled by separate pairs of tubes) produces magnetism in the same single transformer core, the secondary output voltage wave is like that of a three-tube rectifier.* Three tubes $(+A, +B, +C)$ act as a rectifier to produce $(+)$ voltage (shown above the 0 line in Fig. 11F); the other three $(-A, -B, -C)$ act like a separate three-tube rectifier to produce $(-)$ voltage.

While producing a 12-cycle welding current, Fig. 11F shows the order of firing of the six ignitrons, using tube numbers shown also in Fig. 11G. As described later, all the timing of this firing pattern is done by small tubes connected in phase A; these timing circuits control directly only those half cycles shaded in Fig. 11F. The

* Chute, "Electronics in Industry," Chap. 18.

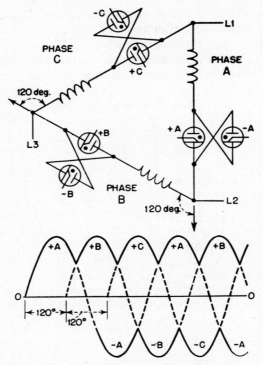

Fig. 11E. Ignitrons in three-phase arrangement.

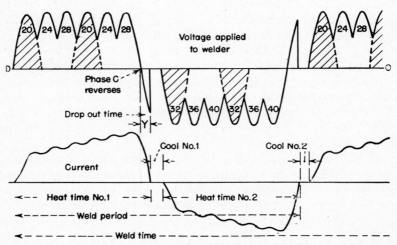

Fig. 11F. Order of firing of ignitrons to supply low-frequency welding current.

tubes in phases B and C are made to fire by the trailing or exciter circuits; whenever the timing circuit fires ignitron 20, the trailing circuits immediately fire ignitrons 24 and 28.*

11-4. General Frequency-changer Operation. Figure $11G$ shows the combined circuit arrangement. As in Fig. $11E$, the 60-cycle three-phase power passes directly to the triangular network of ignitrons and transformer windings; control power connects to a similar triangle of control transformers $35T$, $36T$ and $37T$. Transformer $35T$ supplies phase A (shown later in Fig. $11K$) including control transformers $1T$ and $34T$; $S1T$ and $S34T$ furnish power to the sequence and coordinator panels above.

The sequence timer (tubes 1 to 8) is nearly the same as in Fig. $4H$, explained in Chap. 4. It brings the welder electrodes together during squeeze time, then its tube 4 fires during the weld time, which is the total length of time within which any welding current may flow. During this weld time, there are probably two or more heat times (low-frequency half cycles,† as shown in Fig. $11F$). Within this weld time, the coordinator circuits control the separate heat and cool times. During each heat time 1, tube 17 is firing; during each heat time 2, tube 29 fires. After the weld time ends, the sequence timer continues with the hold time and ends the weld operation.

In the coordinator circuit (upper right in Fig. $11G$), whenever tube 17 fires, transformer $17T$ causes ignitron 20 below to fire in phase A (as described in detail in Sec. 11-8). Another $S17T$ winding fires tube 21 in the exciter-B panel; transformer $21T$ causes ignitron 24 to fire in phase B. Another $S21T$ winding fires tube 25 in the exciter-C panel; transformer $25T$ causes ignitron 28 to fire in phase C. We see that tube 17 indirectly fires ignitrons

---

* At Y in Fig. $11F$, notice that current flows in the welding transformer and ignitron 28 for a short time after the phase-C voltage reverses. This delay Y (called "drop-out time") decreases the cool-time gap between Heat 1 and Heat 2; it varies with the design of welding machine. At lower machine power factors this gap may disappear; the cool time then should be increased to prevent fault current through the ignitrons. If the welding machine uses a transformer having a very large iron core (as is needed for making large welds), the required cool time may exceed 2 or 3 cycles. This may be decreased by adding inversion control, as is described in Sec. 11-12.

† In Fig. $11F$, for example, if the weld time is ¼ sec, welding current flows during 3 cycles of the low-frequency (12 cps) supply. This welding current will include six pulses, or heat times, separated by five cool times. After each negative pulse of low-frequency current, tube 4 stops for an instant as described later.

162   ELECTRONIC MOTOR AND WELDER CONTROLS

20, 24 and 28 during heat time 1. Similarly, tube 29 indirectly fires ignitrons 32, 36 and 40 during heat time 2.*

To study these circuits in detail, first let us describe the sequence and coordinator sections that produce the alternate firing of tubes 17 and 29; later we will confirm that tube 17 represents the firing of all (+) ignitrons, while tube 29 represents the (−) ignitrons.

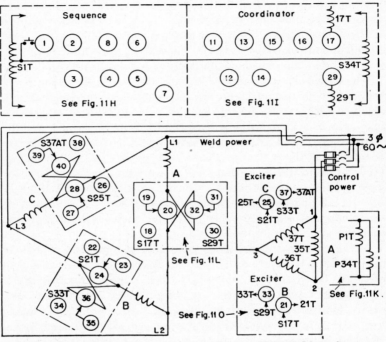

Fig. 11G. Key diagram of circuits and tubes in CR7503–M101 frequency-changer control.

11-5. Sequence Timer of Frequency Changer. The upper section of the control equipment shown in Fig. 11C is the sequence timer; its diagram appears in Fig. 11H. This timer is like that described in Sec. 4–1 except† that there is no WTD relay or no-weld

* Whenever tube 29 fires, transformer $29T$ causes ignitron 32 to fire in phase A. Another $S29T$ winding fires tube 33 in the exciter-B panel; transformer $33T$ causes ignitron 36 to fire in phase B. Another $S33T$ winding fires tube 37 in the exciter-C panel; transformer $37AT$ causes ignitron 40 to fire in phase C. These actions are described in Secs. 11–9 and 11–10.

† Figure 11H omits many details shown in Fig. 4H. However, these timers have identical circuits except for details described above.

switch; instead, tube 4 applies voltage to transformer $4T$ whose secondary $S4T$ controls the shield grid of tube 6. Wires from tube-4 anode 43 and grid 46 connect to the coordinator in Fig. $11I$; there tube 15 returns a signal by wire 63 to the tube-6 anode.

Before the starting switch closes, no tubes are firing in Fig. $11H$ except that tube 8 passes the small currents to charge capacitors $21C$ and $51C$.* As described earlier,[4-3] when the starting switch closes, tubes 1 and 2 fire (in half cycles A, wherein $70A$ is more positive than center line 10), causing relay $SVTD$ and the solenoid valve to bring the electrodes together. Since tube-2 anode 23 now remains 15 volts above cathode 10, tube 3 fires in the following half cycle B; transformer $3T$ keeps tube 2 firing even if the starting switch is released. During the squeeze time, $21C$ discharges until its terminal 26 rises close enough to cathode 10 to fire tubes 4 and 5.

Welding current starts when tube 4 fires (if the no-weld switch is closed, in Fig. $11K$), for tube 4 fires tube 17 in the coordinator and the (+) ignitrons. In detail, during half cycles B (wherein $70B$ is more positive than 10) the tube-4 anode 43 is held about 15 volts above cathode 10; since $P4T$ is an inductive load,[4-4] point 43 remains positive well into the following half cycle A; this fires tubes 11 and 17 in Fig. $11I$, as described in Sec. 11–7. Meanwhile, with tube 5 firing also, $51C$ is discharging at the rate set by weld-time adjuster $51R$. At the end of this total weld time (see Fig. $11F$), the $51C$ terminal 59 has risen close enough to cathode 10 to let tube 6 fire. However, since the tube-6 shield grid is held negative by $S4T$ so long as tube 4 fires, tube 6 waits until tube 4 stops firing for an instant (which occurs just before the end of each negative half wave of low frequency). When tube 6 fires, energizing transformer $6T$, the $S6T$ voltage drives negative the tube-4 shield grid 46 so that tube 4 cannot start another heat-time flow of welding current; this same turn-off signal is carried by wire 46 to the coordinator, to turn off tubes 12 and 14 as is described later. Since tube-6 anode 63 now remains 15 volts above cathode 10, $61C$ discharges at a rate set by hold-time adjuster $61R$; then terminal 65 has risen high enough to fire tube 7. This applies voltage to $P7T$ and also charges $71C$, nega-

---

* Hold-time capacitor $61C$ is not charged by tube 8 (as permitted by WTD contacts in Fig. $4H$); $61C$ is charged by grid rectification in tube 7. These electrons flow from center line 10 through $P7T$, from tube-7 cathode to grid 74, through $74R$ to $61C$, continuing through $P6T$ to $70A$.

164   ELECTRONIC MOTOR AND WELDER CONTROLS

tive at terminal 14. This 71C voltage drives negative the tube-1 control grid; tube 1 drops out relay *SVTD* and the electrodes separate. If the switch is set for a repeat weld, and the starting switch still is closed, all tubes stop firing, letting 71C discharge through off-time adjuster 71R, after which tubes 1 and 2 refire, starting another weld operation.

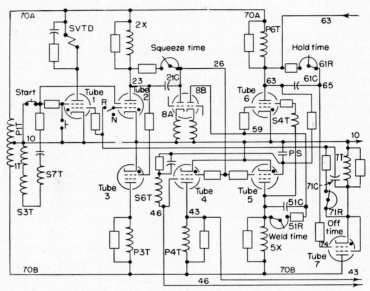

Fig. 11H. Circuit of sequence-timer of frequency-changer control.

Now let us see what takes place in the coordinator section whenever sequence tube 4 fires.

11–6. Coordinator Action. As shown in Fig. 11I, the coordinator section includes heat-time selector 2S and cool-time selector 3S.* Let us use the settings (with 2S and 3S closed to give shortest heat and cool times) that produce the 12-cycle weld frequency shown in Fig. 11F. As is shown in Fig. 11J and explained below,

---

* Switches 2S and 3S are set to give best results with the welding machine, and are not to be changed by the operator. When 2S is opened, Heat 1 and Heat 2 both are increased 50 per cent so that each has a length of 3 cycles (of the 60-cycle supply); operating on three-phase supply, each heat time now includes 9 voltage peaks; the weld frequency becomes 8$\frac{4}{7}$ cycles per second. Similarly, 3S may be set to increase both the Cool 1 and Cool 2 by 1 cycle or by 2 cycles, with further reduction in weld frequency.

each time that tube 4 fires in Fig. 11H, tubes 11 and 17 in Fig. 11I start a (+) wave of welding current. Tube 12 fires after heat time 1, stopping tube 17. Tube 13 fires after cool time 1, followed at once by tube 29, starting a (−) half cycle of welding current. Tube 14 fires during heat time 2, followed by tube 15. Tube 15 applies voltage to transformer 6T in Fig. 11H; this turns off tube 4,

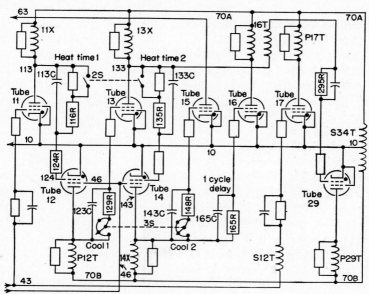

Fig. 11I. Circuit of coordinator portion of frequency-changer control.

and also tubes 12 and 14, followed by tube 13. These actions are timed so that they turn off tube 29 to end heat time 2. If the total weld time (set by 51R) has not yet finished, to let tube 6 fire, then tube 4 will refire to start another weld period [including both a (+) and a (−) half cycle of low-frequency welding current].

Tube 16 has no action in this operation. But if 3S is adjusted to give longer cool time, then tube 15 fires too late to shut off tubes 13 and 29; tube 16 will fire (at X in Fig. 11J) one cycle later than tube 14 and will turn off tube 29 at Y.

11-7. Coordinator in Detail. Until tube 4 fires (in the sequence timer of Fig. 11H), no tube is firing in the coordinator, Fig. 11I. All control power is received from transformer 34T, at the right. Tubes 11, 13, 15, 16 and 17 receive positive anode voltage only during half cycles A (wherein 70A is more positive than cathode

10); during such half cycles their control grids, being at 70B potential, are more negative than 10. Similarly tubes 12, 14 and 29 receive positive anode voltage only during half cycles B, wherein 70B is more positive than cathode 10; their control grids, being at 70A potential, are negative during these B half cycles. The shield grids 46 of tubes 12 and 14 are at the same potential as the tube-4 shield grid; these are at cathode-10 potential after pressure switch PS has closed, in Fig. 11H.

Near tube 11, capacitor 113C is charged by grid rectification of tube 12; during half cycles A electrons flow from cathode 10 to tube-12 control grid 124, through 124R to 113C, through 11X to 70A. Similarly, 133C is charged through tube 14. During half cycles B, 123C, 143C and 165C are charged through tubes 13, 15 and 16.

To make a weld, at the end of squeeze time tube 4 fires in a half cycle B. Its anode 43 now remains about 15 volts (arc drop) more positive than cathode 10. Because of "inductive hangover"[4-4] this potential at 43 fires tubes 11 and 17 in half cycle A (above the 0 line in Fig. 11J). Tube 17 applies voltage to $P17T$ to fire the (+) ignitrons as is described later. When tube 11 fires, its anode 113 remains 15 volts above cathode 10; the charge on 113C holds negative the tube-12 grid during the following cycle. When 113C has discharged through heat-time resistor 116R enough to fire tube 12, tube 17 already has fired twice (as shown at U in Fig. 11J), so that the (+) ignitrons produce the positive welding-current wave lasting two cycles (of the 60-cycle supply). However, tube 12 applies voltage to $P12T$ whose secondary drives negative the tube-17 grid to end the positive welding-current wave.*

Since the cool time is less than $\frac{1}{3}$ cycle, 123C discharges through 129R so quickly that tube 12 fires tube 13 in the following half cycle A, as shown at V. The tube-13 anode 133 remains 15 volts above 10; since 133 connects through $S16T$ and 295R to the tube-29 grid, tube 29 fires in the following half cycle B and applies voltage to $P29T$ to fire the (−) ignitrons. Near tube 13, 133C discharges so quickly through heat-time resistor 135R that tube 14 also fires in the half cycle B following tube 13. With tube-14 anode 143

* Although tube 17 does not fire in the third cycle, notice (in Fig. 11J) that ignitron 28 in phase C still is passing current far into the third cycle. Less than $\frac{1}{3}$ cycle remains between the ending of ignitron-28 current and the starting of ignitron-32 current in the following half cycle B.

being held 15 volts above 10, 143C discharges so quickly through cool-time resistor 148R that tube 15 is fired in the following half cycle A. In this same half cycle A (as shown at W in Fig. 11J) notice that tube 13 has been fired a second time by tube 12, so tube 29 likewise is fired a second time to produce a negative wave of welding current that is 2 cycles long. However, the firing of

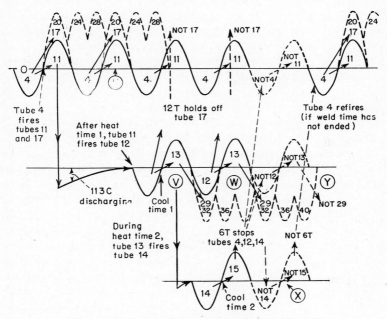

Fig. 11J. Action of tubes in coordinator, Fig. 11I.

tube 15 applies voltage to P6T in Fig. 11H. Although tube 6 is not firing at this time, tube 15 makes S6T (near tube 4) produce a voltage to drive negative the shield grid 46 of tube 4 and also the shield grids of tubes 12 and 14. Since 6T thus prevents tubes 4, 12 and 14 from firing, there is no "inductive-hangover" grid potential to fire tubes 11, 13 or 15 in the following half cycle; therefore tube 13 cannot fire tube 29 a third time (at Y), so the negative wave of welding current is ended.

During this half cycle when tube 4 is held off by S6T, the S4T voltage disappears from the tube-6 shield-grid circuit. Tube 5 is firing during this time, so the tube-6 shield grid now is at the +15-volt potential of tube-5 anode 53. Therefore, if adjuster

168   ELECTRONIC MOTOR AND WELDER CONTROLS

51R is set for such a short weld time that 51C already has discharged, control grid 59 has reached high enough potential to fire tube 6. If this happens, tube 6 applies voltage to $P6T$ (in place of tube 15); $S6T$ prevents tube 4 from further firing. The sequence timer completes the hold time (as described in Sec. 11–5) and ends the welding operation.

If 51R is set for a weld time longer than one complete low-frequency cycle of welding current, the potential at control grid 59 prevents tube 6 from firing yet. At X in Fig. 11J it is seen that tube 15 does not fire more than one half cycle; therefore $S6T$ does not hold off tube 4 more than one half cycle. Tube 2 still is firing, keeping the tube-4 control grid positive. So, after missing one half cycle, tube 4 again fires tubes 11 and 17, starting the next positive half wave of low-frequency welding current. The refiring of tube 4 also causes $S4T$ to prevent tube 6 from firing until tube 4 again is turned off, near the end of the next low-frequency cycle.

11–8. Phase-A Exciter. Auxiliaries. The circuits described above receive their power through control transformers 1T and 34T, whose primary windings appear at the right-hand side in Fig. 11K. All parts of Fig. 11K work in phase A, and receive power through transformer 35T (connected correctly for the supply voltage between lines 1 and 2). Here are the protective devices for the entire frequency-changer equipment. When control power has been applied for 5 minutes to warm the tubes, time-delay relay TD picks up relay CR which seals around TD. Meanwhile cooling water must flow in the ignitrons to hold closed the thermal-flow-switch contacts TFS. If the no-weld switch is closed, the CR contact picks up three relays; 1CR closes the anode circuits of thyratrons 19 and 31 in phase A (as shown in Fig. 11L). Similarly, 2CR controls tubes 23 and 35 in phase B; 3CR controls tubes 27 and 39 in phase C. Unless all ignitrons have cooling water and the other tubes are warmed, no ignitron may fire; there is no welding current. The red signal light indicates that control voltage is present; amber, that cooling water flows; green, that the 5-minute warming time is ended. When the no-weld switch is closed, welding may proceed.

In Fig. 11K transformers 35 and 41 supply power to heat the tubes in the heat-control and firing circuits of Fig. 11L. Transformer 38 supplies a sine-wave voltage to fire tubes 19 and 31 in Fig. 11L, as explained later. The phase-shifting circuit of $P38T$, $A41C$, $A1R$ and $A4R$ (operating from terminals 6A, 8A and mid-

THREE-PHASE WELDING CONTROLS 169

tap $7A$ of $35T$) is the same as in the CR7503–D175 heat control described in Sec. 1–5 and shown in Fig. $1K$; there the corresponding parts are $P2T$, $4C$, $4R$, $5R$ across $P1T$. The heat-adjusting dial $A4R$ is mounted in the operator's control station; the knob that turns $A4R$ also turns $B4R$ (shown in Fig. $11N$) and $C4R$ in phase C. In this way the ignitrons in all three phases may have the same amount of phase retard at the reduced heat setting.

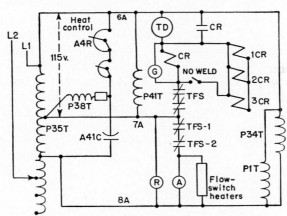

Fig. $11K$. Circuit of phase-A exciter and auxiliaries.

11-9. Firing of Ignitrons. Figure $11L$ includes the tubes that fire directly ignitrons 20 or 32 in phase A. A duplicate circuit fires ignitrons 24 or 36 in phase B; a third circuit fires ignitrons 28 or 40 in phase C. Notice that the left-hand half in Fig. $11L$ is arranged like the right-hand half, as if reflected in a mirror; only one half needs to be explained in detail.

As mentioned previously,[11-4] when tube 17 (in the coordinator, Fig. $11I$) applies voltage to transformer $17T$, the $(+)$ ignitrons fire. In Fig. $11L$, this signal from $S17T$ appears at the left, at the cathode of tube $18A$.* Similarly, when tube 29 fires, $S29T$ produces the signal voltage at tube-$30A$ cathode (at the right), to let $(-)$ ignitron 32 fire.

First let us see what prevents tubes 19 and 20 from firing (at the left in Fig. $11L$) when there is no $S17T$ voltage. Since we know (from Sec. 1–4) that ignitron 20 fires only when thyratron

---

* The dotted sides of the circles of tube $18A$ and $18B$ indicate that these two diodes are within the same single enclosure. Tube 18 is a 6H6 double rectifier tube.

170   ELECTRONIC MOTOR AND WELDER CONTROLS

19 fires first, we must study the grid circuit of tube 19. If we remove tube 18 for a moment, tube 19 fires. The tube-19 grid now is controlled entirely by the a-c voltage of $S38T$; this voltage always is present and fires tube 19 whenever the $1CR$ contact is closed above it.* By turning the heat-control dial ($A4R$ in Fig. 11K), the phase position of the $S38T$ voltage wave may be moved to fire tube 19 earlier or later in the half cycle of phase-A voltage.

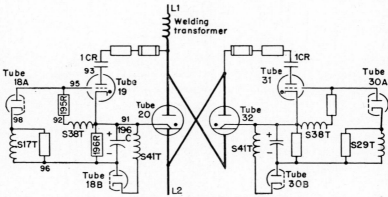

Fig. 11L. Arrangement of tube circuits to fire one pair of ignitrons.

When tube 18 is returned to the circuit, tube 19 stops firing although the $S38T$ voltage still is present. Tube 18B acts as a rectifier to let transformer $S41T$ charge capacitor 196C so that its terminal 96 remains about 60 volts more negative than cathode 91 of tube 19.† This condition is shown also at the left in Fig. 11M. When no voltage is produced by $S17T$, the cathode 98 of tube 18A is at this negative potential of point 96. When the $S38T$ voltage raises its terminal 92 to a potential more positive than cathode 98, tube 18A conducts; electrons flow from 196C negative terminal 96 to cathode 98, through tube 18A to anode 95, through 195R to $S38T$, to positive terminal 91 of 196C. So, although terminal 92 of $S38T$ becomes positive, tube 18A prevents grid 95 from rising high enough to fire tube 19.

Let us see how the firing of tube 17 (in Fig. 11I) causes 17T to fire tube 19. In the second cycle shown in Fig. 11M, the voltage

---

* Without tube 18, this circuit action is like the firing of tubes 1 and 2 in Fig. 1K.

† Since the time constant (196R × 196C) is 1 sec, there is little ripple in this 196C voltage.

THREE-PHASE WELDING CONTROLS 171

wave of $S17T$ overcomes the bias of $196C$, and forces cathode 98 of tube $18A$ to become more positive than cathode 91 of tube 19. Although the tube-$18A$ anode 95 is being raised by $S38T$ as before, the potential at 95 does not rise above cathode 98 until late in the half cycle. Meanwhile grid 95 has fired tube 19; tube $18A$ conducts too late to affect the firing of tube 19.

While $S17T$ is thus letting tube 19 fire ignitron 20, tube 30 is preventing the right-hand $S38T$ winding from firing tube 31. Later, when tube 29 fires in the coordinator (Fig. $11I$), the $S29T$ voltage holds off tube $30A$ until the $S38T$ voltage wave has fired tubes 31 and 32.

11–10. Trailing Exciter Circuits for Phases B and C. While transformers $17T$ and $29T$ thus control the tubes in phase A, the voltage waves of $S17T$ and $S29T$ are not in phase position suitable for the firing-tube circuits of phases B or C. Instead, $S17T$ and $S29T$ voltages are used in the exciter-B circuit shown in Fig. $11O$.* Notice the $S17T$ winding in the grid circuit of tube 21. Tube 21 receives anode voltage from terminal $6B$ of transformer 36 (in phase B from lines $L2$ and $L3$). As shown in Fig. $11N$, the wave of tube-21 anode voltage is 120 deg later than for a phase-A tube (such as tube 19 in Fig. $11M$). When tube 17 is not firing (in the coordinator, Fig. $11I$), $S17T$ produces no voltage, so the tube-21 grid is at the potential of $8B$. Since $8B$ is negative whenever anode $6B$ is positive, tube 21 cannot fire.†

When tube 17 fires, the $S17T$ voltage appears in Fig. $11N$ (in phase-A position, lining up with $S17T$ above in Fig. $11M$). This $S17T$ voltage remains positive in the early part of phase B when tube-21 anode $6B$ has become positive. This fires thyratron 21, which applies voltage to $P21T$ during a half wave of phase-B voltage, even though the tube-21 grid has been driven far negative soon after firing tube 21, as shown at H. Therefore transformer $21T$ supplies voltage ($S21T$ in Fig. $11N$) in the correct phase position to hold off tube $22A$ in the phase-B firing circuit, to let tube 23 fire ignitron 24 (as outlined in Fig. $11G$), in the same way as is described earlier for $S17T$ with tubes $18A$, 19 and 20.

Similarly, when tube 29 fires in the coordinator, the $S29T$

---

* The phase-shifting circuit ($P39T$, $B41C$, $B1R$ and $B4R$) is like that in Fig. $11K$ but is shown upside down, since $P36T$ is located at 120 deg to $P35T$, as shown in Fig. $11G$.

† The voltage across $215C$ (charged by grid rectification) drives grid 214 negative slightly before anode $6B$ becomes positive, as shown at G in Fig. $11N$.

winding in Fig. 11O fires tube 33; transformer 33T produces a wave of phase-B voltage to hold off tube 34A in the phase-B firing circuit, to let tube 35 fire ignitron 36.

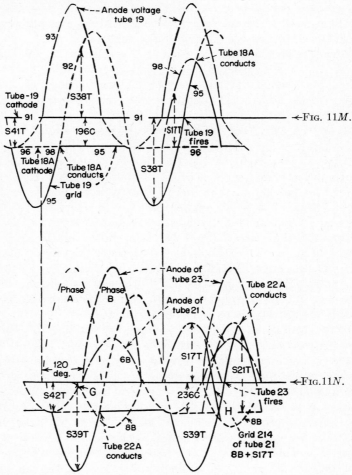

FIG. 11M. Action whereby tube 17 in the coordinator fires tube 19 and ignitron 20.
FIG. 11N. Action in Fig 11O to make the phase-B tubes trail the phase-A tubes.

Since the voltage waves produced by transformers 21T and 33T in phase B (or 17T and 29T in phase A) do not have phase position suitable for the firing-tube circuits of phase C, another exciter circuit is used (shown as exciter C in Fig. 11G) similar to Fig. 11O. Just as an $S17T$ voltage (phase A) is used to fire tube 21 (phase B) as shown in Fig. 11N, the exciter-C circuit uses an $S21T$ voltage (phase B) to fire tube 25 (phase C). Tube 25 applies voltage to transformer 25T, so that $S25T$ produces a wave of phase-C voltage to hold off tube 26A in the phase-C firing circuit, to let tube 27 fire ignitron 28.*

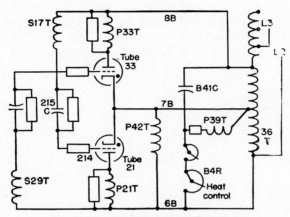

Fig. 11O. Circuit of phase-B exciter.

11–11. Over-all Tube Sequence. In summary, let us note that the various tubes (scattered among the coordinator, exciter and firing sections) are numbered in their normal order of firing. Tubes 1 to 16 supervise the time length and the number of the current waves as described earlier in Secs. 11–4 to 11–6. However, each time tube 4 fires, a full wave of low-frequency welding current is produced by tubes 17 to 40. Tube 4 fires 17 to start the (+) half cycle; 17T holds off 18, so 19 may fire ignitron 20 (+A). In exciter B, 21 fires and 21T holds off 22, so 23 may fire 24 (+B). In exciter C, 25 fires and 25T holds off 26, so 27 may fire ignitron 28 (+C). Back in the coordinator, tube 29 starts the (−) half cycle; 29T holds off 30, so 31 may fire ignitron 32 (−A). In exciter B, 33 fires and 33T holds off 34, so 35 may fire

* Similarly, when tube 29 fires and $S29T$ fires tube 33 (phase B), the exciter-C circuit uses an $S33T$ voltage (phase B) to fire tube 37 (phase C). Tube 37 applies voltage to transformer 37AT, so that $S37AT$ produces a wave of phase-C voltage to hold off tube 38A in the phase-C firing circuit, to let tube 39 fire ignitron 40.

ignitron 36 ($-B$). In exciter C, 37 fires and $37AT$ holds off 38, so 39 may fire ignitron 40 ($-C$).

11–12. Inversion Used for Faster Current Decay. When larger welding transformers are used with the control described above, longer drop-out time may be required (see Fig. 11F and Sec. 11–3, footnote), because of the greater amount of energy stored in the transformer. This energy can force current to flow through the phase-C ignitron long after its anode voltage has reversed. As a result, in part (a) of Fig. 11P, the drop-out time Y is greater than in Fig. 11F. Primary current continues to flow until E, although the shaded negative voltage is opposing this current.

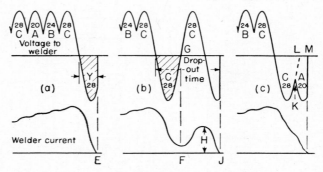

Fig. 11P. Showing the need for inversion control.

A more extreme case is shown in part (b); here the entire shaded half cycle of voltage is not able to reduce the welder's current to zero. At F, some current still flows in ignitron 28 when its anode voltage returns positive at G; this ignitron remains ionized and continues to fire (although it receives no ignitor current from its thyratron 27) and applies positive voltage to increase the current as is shown at H. The current finally stops at J, but this long drop-out time is undesirable.

To decrease this drop-out time, part (c) shows that ignitron 20 in phase A may be fired (by the inversion* control described below) so as to continue to apply negative voltage to decrease the

---

* Inversion describes the action of a tube when its grid (or its ignitor) purposely is controlled at the proper instant so as to make the tube continue to conduct current after the line voltage has reversed; during such inversion, the transformer's stored energy causes power to flow through the ignitron into the supply line. Inversion occurs in the phase-C ignitron during Y in Fig. 11P. Additional inversion occurs when the phase-A ignitron 20 is fired by this accessory control.

welder current. At point K the current becomes less in phase-C tube 28 and increases in phase-A tube 20, until tube 28 no longer is ionized when its anode voltage returns positive at L. The negative voltage applied through tube 20 further decreases the welder current until it becomes zero at M.

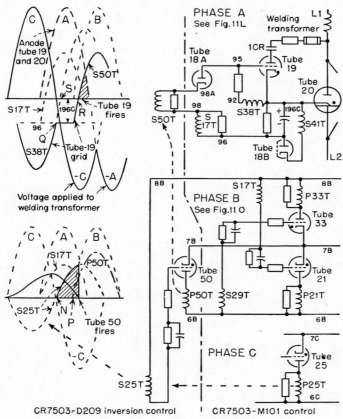

Fig. 11Q. Circuit and action of inversion-control accessory.

11–13. Inversion-control Accessory (CR7503–D209). To fire ignitron 20 at the right time so as to shorten the positive flow of welder current as described above, accessory circuits are added (to the frequency-changer welder control of Sec. 11–3) as shown in Fig. 11Q.* To fire ignitron 20 in phase A, a transformer winding $S50T$

---

* Similarly to shorten a negative half cycle, the circuit for firing ignitron 32 in Fig. 11L is provided with a corresponding inversion circuit like that shown in Fig. 11Q.

is inserted below tube 18A. The primary winding $P50T$ receives voltage from phase B whenever tube 50 fires. Tube 50 is fired by the transformer winding $S25T$, whose primary winding $P25T$ receives voltage from phase C while tube 25 is firing.

Normally ignitron 28 in phase C is the last tube to pass current to the welding transformer in the low-frequency positive half cycle (see Fig. 11F). Figure 11G shows that ignitron 28 is fired when tube 25 applies phase-C voltage to 25T, in the exciter-C circuit. With inversion control added, tube 25 also causes $S25T$ to fire tube 50 so that tube 19 fires ignitron 20 in phase A; this delayed firing of ignitron 20, following after ignitron 28 of phase C, reduces the welder current to zero more quickly as shown in (c) of Fig. 11P.

Before tube 25 fires in phase C (at lower right in Fig. 11Q), there is no voltage from $S25T$, so the grid of tube 50 is held negative at the potential of point 8B. When tube 25 fires during a weld, Fig. 11Q shows at N (at lower left) how the $S25T$ voltage (in phase C but of reversed polarity) becomes positive after the phase-C voltage has reversed; in this way the grid of tube 50 is held positive until after the phase-B voltage drives point 6B positive at P, letting thyratron tube 50 fire.

At the upper left, Fig. 11Q shows the usual position of $S17T$. However, there is no $S17T$ voltage to refire the phase-A ignitron, for tube 17 in the coordinator (Fig. 11G) has stopped firing, to end heat time 1. Therefore tube 18A conducts at Q, as described in Sec. 11–9 and Fig. 11M, and holds the tube-19 grid at the negative potential of point 96. However, at R, the voltage of $S50T$ is produced by the firing of tube 50 in the inversion control; this voltage raises cathode 98A of tube 18A. Since the $S38T$ voltage previously has raised its terminal 92 positive at S, tube 19 fires. In this way, ignitron 20 is fired by tube 19 just before the phase-A voltage becomes negative.

PART II
MOTOR CONTROLS

CHAPTER 12

POSITION CONTROLS

The first part of this book describes controls for resistance welding; as such, the electronic circuits include thyratrons and ignitrons with split-second timing, so that large currents produce the exact amount of heat needed to make good welds. In contrast, this second part describes electronic circuits that control motors; here the main purpose is to produce motion and to control its speed and force more accurately or more quickly than was possible with earlier nonelectronic equipment. Here we shall find many small high-vacuum tubes used to operate relays, or to "tell" thyratrons or amplidynes how to control motor speed.

12-1. Motor Controls. This chapter describes several types of electronic equipment that do not try to control the speed of a motor; they merely operate relays that "tell" a motor which way to turn, at full speed, to move an object or a load into a desired position. Later chapters will describe d-c motors that receive part or all of their direct current through electron tubes, fed from the a-c supply lines; such tubes accurately may control the speed of such a motor through a wide range or they may set the amount of force or tension that the motor causes.

12-2. Electronic Positioning Control (CR7510–A102). This equipment, shown in Fig. 12A, is a general-purpose follow-up control; that is, it may be used with any type of motor* to move an object into any position selected by turning a small dial. The equipment includes the control panel and two potentiometers; for increased accuracy the panel may be used with two selsyns.

Figure 12B shows the diagram of this electronic control used with two potentiometers A and B and a reversing motor, geared to drive a load.

As an example, if you wish to control a motor-operated valve

---

* A small motor (up to $1/16$ hp) may operate directly from the relay contacts; for motors to $1\frac{1}{2}$ hp, these relays may control the contactors of a reversing switch. Larger motors may require further study of the system controlled.

from a distance, you may turn the potentiometer slider A to any point on its dial; immediately, perhaps 500 ft distant, the motor will turn at full speed until it has brought the valve to the position that matches your setting of slider A. The second potentiometer B is connected to the valve; the motor drives the valve until the slider on potentiometer B reaches a position that matches the position of slider A.

FIG. 12A. Electronic positioning control (CR7510–A102), with selsyns.

Briefly, when no action is desired in Fig. 12B, and slider B is in the same position as slider A, transformer 1T has no voltage, so both tubes 1A and 1B are passing current; neither tube 2A nor 2B is passing enough current to pick up its relay. If slider A is turned to the right (clockwise), this turns off tube 1A so that tube 2A passes current and picks up relay 1CR; the motor turns forward so that slider B also moves to the right. However, if slider A is turned to the left, this turns off tube 1B so that tube 2B picks up 2CR; the motor is reversed so that slider B is moved to the left.

All power for the tubes (including their heaters, not shown) and relays is supplied through transformer 2T at the left in Fig. 12B. Although four tube circles are shown, only two separate tubes are used, each being a twin triode (like the tubes in Fig. 12I). A-c voltages are applied across the tubes; however, no tube can pass current during the negative half cycles when top point 18 is more negative than cathodes 17 or 23. We need to study only what happens during the positive half cycles, when transformer terminal 18 is more positive than 17. During such half cycles, note that terminal 25 of potentiometer 1P is 4 volts more negative than point 17. When transformer 1T is producing no voltage

POSITION CONTROLS 181

(so that no voltage appears across 1R, 2R, 1C or 2C), the grid voltage of tubes 1A and 1B may be varied between −4 and +14 volts, depending entirely on the setting of 1P. Even when 1P is set so that the grids are negative, tubes 1A and 1B pass a tiny current which, flowing through the high resistance of 3R or 4R,

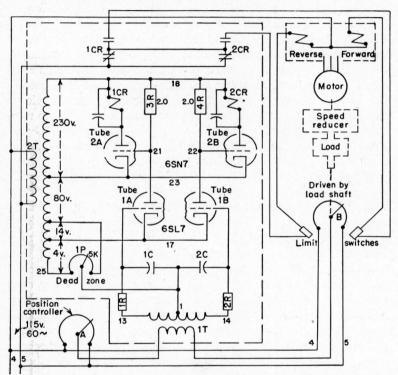

Fig. 12B. Circuit of electronic positioning control (CR7510–A102).

causes a large voltage drop across these resistors. Grids 21 and 22 are held far more negative than cathode 23, so tubes 2A and 2B do not pick up relays 1CR or 2CR. This condition continues so long as slider A is not turned away from the same position as slider B.

Potentiometers A and B are connected to the same 115-volt a-c supply. While their sliders are in the same position, the a-c voltage from line 4 to slider A is the same as the a-c voltage from line 4 to slider B, so the two sliders are at the same potential and no voltage is applied to transformer 1T.

182   ELECTRONIC MOTOR AND WELDER CONTROLS

12–3. Follow-up Action. First let us set the $1P$ slider to touch near point 25 in Fig. $12B$. As is shown in Fig. $12C$, the potential at point 1 (slider of $1P$) now is close to the cutoff potential of tube 1.* If slider A now is turned a small amount to the right, the voltage from line 4 to slider A becomes slightly greater than the voltage from line 4 to slider B. This small voltage difference (often called the error voltage) is applied to transformer $1T$; during the positive half cycles, the secondary voltage of $1T$ now forces its terminal 13 more negative than center tap 1, but terminal 14 becomes more positive than 1. Figure $12C$ shows that this small signal voltage from $1T$ forces the tube-$1A$ grid closer to cutoff;† the current decreases through tube $1A$ and $3R$, letting grid 21 rise higher so that the tube-$2A$ current increases to pick up relay $1CR$.‡ Meanwhile the $1T$ signal voltage raises the tube-$1B$ grid, lowering further the grid-22 potential; this has no effect on relay $2CR$, which remains deenergized.

When $1CR$ is picked up, a circuit is completed to the forward contactor coil of the reversing switch. The motor drives the valve load in the forward direction and this also moves slider B to the right. This movement of slider B decreases the error voltage applied to $1T$ until the tube-$1A$ current increases and relay $1CR$ drops out.

If slider A now is turned to the left, a voltage difference (or error voltage) again is applied to transformer $1T$. However, since the voltage from line 4 to slider A now is less than the voltage from line 4 to slider B, the small voltage applied to $1T$ has opposite phase relation to the previous error voltage (or is 180 deg out of phase). So, as is shown in Fig. $12D$, the grid of tube $1A$ becomes more positive and relay $1CR$ is not affected; the grid of tube $1B$

---

* Cutoff is that grid voltage at which the anode current is zero. At a higher anode voltage, the grid must be made still more negative to cut off the anode current.

† The potentiometer or selsyn system causes a phase shift between the $1T$ signal voltage and the line voltage (or anode voltage of the tubes). Capacitor $1C$ acts with resistor $1R$, so that the voltage across $1C$ is brought in phase; this $1C$ voltage is the signal actually used to control tube $1A$, as shown in Fig. $12C$.

‡ Here tubes 1 and 2 serve as a two-stage amplifier. Turning slider A through half a degree causes only $\frac{1}{6}$ volt across $1T$. This tiny voltage at the grid changes the tube-$1A$ current enough to vary the point-21 potential by 6 or 8 volts; the tube-$2A$ current increases by 10 milliamperes, enough to pick up $1CR$.

is forced more negative, decreasing the tube-$1B$ current and raising grid 22 so that tube $2B$ picks up $2CR$. Relay $2CR$ completes the circuit to the reverse contactor coil, so the motor drives the valve load in the reverse direction and moves slider B to the left until the error voltage at $1T$ decreases and $2CR$ drops out.

12–4. Accuracy vs. Dead Zone. When the dead-zone adjuster $1P$ is set so that its slider touches at point 25 (in Fig. 12B), a very small movement of slider A (perhaps only ¼ degree) may pick up relay $1CR$ or $2CR$; the motor turns, trying to position the load (such as a valve) and slider B to within this ¼-deg accuracy.

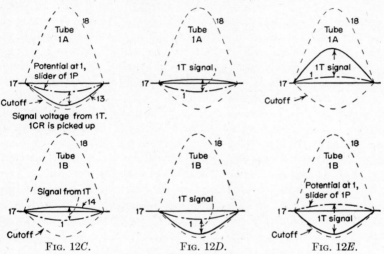

Fig. 12C. Fig. 12D. Fig. 12E.

Fig. 12C. Action in Fig. 12B as slider A is turned to the right.
Fig. 12D. Action in Fig. 12B as slider A is turned to the left.
Fig. 12E. The dead zone is increased as compared with Fig. 12D.

However, such close follow-up or accuracy can rarely be used, unless the motor is stopped very quickly (as by a brake on the motor shaft). When the motor brings slider B to the position where relay $1CR$ drops out (thereby removing power from the motor), the motor and its load may coast far enough to move slider B beyond slider A so that $2CR$ picks up, reversing the motor. This may cause the motor to turn back and forth, although slider A is not being moved; this action is called "hunting," and must be stopped.

If $1P$ now is turned away from 25, raising the point-1 potential as shown in Fig. 12E, slider A must be turned a greater amount (such as 2 deg) before the signal voltage of $1T$ is large enough to force the tube-$1B$ grid low enough to make tube $2B$ pick up $2CR$. Similarly, as the motor drives slider B toward the same position as slider A, relay $2CR$ may drop out while slider B still is 2 deg out of line. This permits the coasting motor to drive the load and slider B through an extra 2 or 3 deg without causing tube $1B$ to pick up $1CR$. So, by turning $1P$ slider away from 25, a larger neutral or dead zone is produced within which no correction occurs. If the motor and load coast far enough to move slider B through 3 deg after $2CR$ drops out, then $1P$ must be set to provide a dead zone of nearly 4 deg; this limits the accuracy to plus or minus 2 deg, since slider A may be turned 2 deg before $1CR$ or $2CR$ operates. We shall see that the accuracy of any follow-up system depends on the width of the neutral or dead zone needed to prevent hunting.

12–5. Closed-cycle Systems. The positioning control described above is an example of a closed-cycle system; by turning slider A you produce an error signal that causes a motor to turn in such a direction that slider B is moved to decrease the error signal, bringing the system back into balance. A motor-driven pump that fills a water tank is not a closed-cycle system unless a float or pressure switch is used to send back a signal that stops the motor when the tank is full, or restarts the motor as the water level falls. The welder controls in earlier chapters are open-cycle systems; the welding current is carefully timed and adjusted (by the operator) to produce good welds, but there is no return signal that warns the operator or that changes the current or time settings if the weld quality changes.

All motor controls described here are closed-cycle systems. If constant motor speed is desired, a speed signal is produced at the motor and is sent back to the control circuit to show whether the speed need be corrected. If constant load or tension is desired, a load signal is similarly returned to the control. So, in a positioning control, an electrical signal is produced (as at slider B above) that "tells" the control circuit when the load or object reaches the desired position. Let us now see how a pair of selsyns can produce such a position signal; then a phototube will be described as it "reports back" on the position of a paper roll.

12–6. Selsyn Control of Positioning. In place of the two potentiometers A and B shown at the bottom of Fig. 12B, two selsyns* may be used as shown in Fig. 12F. Three wires connect

* Chute, "Electronics in Industry," Chap. 28.

the stationary parts of the two selsyns. The rotor of selsyn A may be turned through 180 deg by the position-controller dial; the rotor of selsyn B is driven by the load shaft. When the rotor of B is exactly in line with the rotor of A (as indicated by the arrows in Fig. 12F), no voltage is produced from selsyn B to transformer 1T, so the motor does not move the load. If selsyn A is turned to the right, selsyn B produces an a-c voltage at 1T that may

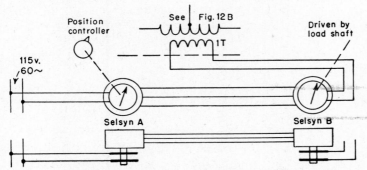

Fig. 12F. Position control by selsyns, combined with Fig. 12B.

turn off tube 1A and pick up 1CR (in Fig. 12B); the motor drives selsyn B to the right, decreasing this 1T voltage. When selsyn A is turned to the left, the voltage produced by selsyn B is of opposite phase and turns off tube 2A, picking up 2CR; the motor drives selsyn B to the left until 2CR drops out.

12-7. Two-point Side-register Control (CR7505–S118). Figure 12G shows a light-operated relay; its phototube (in a separate scanning head, not shown) "looks" at the edge of a continuous web or strip of paper or metal while the strip is being wound onto a roll; it "sees" any small sidewise movement of the strip and makes a motor move the whole roll lengthwise. This is a control for positioning the roll so that its wound edge will be smooth.*

In the circuit of Fig. 12I, phototube 1 responds to a beam of light that is half cut off by the edge of the strip being wound. If this edge moves slowly to the right, cutting off more of the beam, less light reaches the phototube. As is explained below, current decreases in tube 2A, increases in tube 2B, and relay 1CR is picked

* This S118 unit is used when the edge of the strip does not change position more than 10 inches per minute. To control a faster runout or side motion, the S120 or S127 side-register controls are used, as described in Secs. 14–4 and 14–5.

186 ELECTRONIC MOTOR AND WELDER CONTROLS

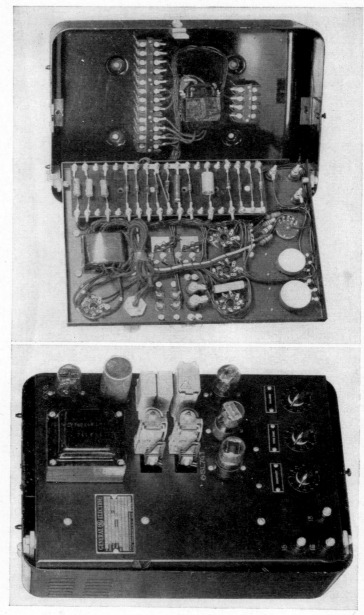

Fig. 12G. Two-point side-register control (CR7505–S118).

up. The 1CR contacts (arranged like those at the top in Fig. 12B, or connected to any type of reversing motor) make the motor move the roll to the left until half of the light beam again reaches phototube 1. If the roll moves too far to the left, the increased light at phototube 1 decreases the tube-2B current so that relay 2CR is picked up and the motor moves the roll to the right.

At the left in Fig. 12I, the a-c control power is rectified by tube 5 and filtered by 1C, 15R and 2C so that 360 volts d.c. is supplied to the other circuits between top point 25 and bottom point 20. By means of the voltage divider 7R, 6R, 5R, 4R, 3R and 8R, only 75 volts appear across phototube 1 in series with resistors selected by the sensitivity switch S. Notice that the potential at point 15 below the phototube gives the grid signal to tube 2A.

For high sensitivity, switch S connects the phototube to high resistance 1R. When very little light reaches phototube 1, little current flows through 1R; grid 15 is at a low potential, close to ground point 13. However, as more light reaches phototube 1, more current flows through 1R so that the potential rises at grid 15 to turn on tube 2A. Here a small change of light causes enough change in phototube current so as to operate the correcting relays 1CR or 2CR. To make the equipment less sensitive, so that the edge of the rolled strip must move sidewise further before 1CR or 2CR will operate, switch S is turned away from 1R, connecting instead to a lower resistance such as 17R. Now the light reaching phototube 1 must change a greater amount to produce the voltage variation across 17R that is needed to operate the correcting relays.

As the strip edge moves, changing the amount of light reaching the phototube, this gives only one signal or grid potential to tube 2A. Let us see how tubes 2A and 2B act as a "long-tailed pair" to change this one signal into two opposing output signals to operate the two relays.

12-8. The "Long-tailed Pair." In Fig. 12H we see two vacuum triode tubes 2A and 2B (usually exactly alike and in the same enclosure); they have equal anode resistors 12R and 13R but their cathodes connect to a single resistor 9R. Any current through either triode must pass through cathode resistor 9R, which is the "long tail" for this pair of tubes. This tube circuit appears also in the light-controlled equipment of Fig. 12I described above. Notice that the grid potential of tube 2B does not change (after it is set by 3R in Fig. 12I).

The long-tailed pair of tubes receives just one grid signal. If this tube-2A grid potential rises at point 15, we will see that tube-2A anode 37 decreases in potential, while tube-2B anode 38 rises in potential. In this way a single input signal is changed into two equal and opposite output signals.

Whenever the circuit of Fig. 12H is in balance, so that the input signal holds grid 15 at exactly the same potential as the steady grid 18, each triode has the same grid-to-cathode voltage; each triode passes the same amount of current, so the voltage across 12R is equal to the voltage across 13R. Point 37 is at the same potential as point 38. If the combined anode currents total 0.6 ma, this current causes 48 volts drop across the 80,000 ohms

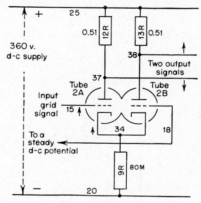

Fig. 12H. Circuit of a "long-tailed pair."

of cathode resistor 9R, so that the cathode 34 is 48 volts above point 20. If both grids 15 and 18 are 46 volts above point 20, the grid voltage (grid to cathode) is −2 volts.

If now the grid signal rises at grid 15 (becomes more positive, as when phototube 1 in Fig. 12I receives more light), this small rise of grid-15 potential will increase the current flowing in 12R and tube 2A, so that the anode potential 37 drops. This greater current also increases the voltage across 9R, so that cathode 34 of tube 2B rises slightly, perhaps to 48½ volts above point 20. Since the potential of grid 18 remains at 46 volts above 20, the grid voltage of tube 2B becomes −2½ volts instead of −2 volts, thus decreasing the current in tube 2B and 13R. Since less current decreases the voltage drop across 13R, the potential at anode 38 rises. The large current increase in tube 2A is nearly offset by the large current decrease in tube 2B, so the total current in 9R has increased very little; this long-tailed pair draws nearly constant current from the d-c supply.

Referring to Fig. 12I, we see that anodes 37 and 38 are connected to the grids of triodes 4A and 4B. When the strip being wound is in the desired central position (so that about half the light beam reaches the phototube) neither tube 4A nor 4B passes enough current to pick up relay 1CR or 2CR. (This is adjusted by 6R and tube 3 as described in Sec. 12–9.) If the strip edge moves to the left, phototube 1 receives more light and point 15

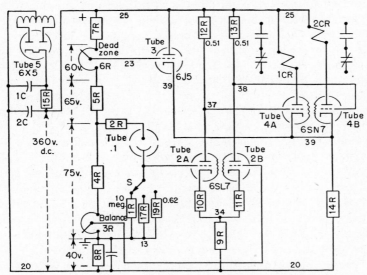

Fig. 12I. Circuit of two-point side-register control (CR7505–S118).

rises, increasing the tube-2A current, then point 37 falls and decreases the tube-4A current to zero; at the same time, the tube-2B current decreases because of the long-tailed-pair action described above, and point 38 rises increasing the tube-4B current to pick up 2CR. Relay 2CR makes a motor move the wound roll lengthwise to the right, to keep the roll edge smooth, also increasing the light at phototube 1. Similarly, when the strip edge moves to the right of the desired central position, the light decreases at phototube 1 and point 15 falls, decreasing the tube-2A current; point 37 rises, increasing the tube-4A current so as to pick up relay 1CR and move the wound roll toward the left. Meanwhile, in the long-tailed pair, the decrease of tube-$2A$ current also decreases the current through $9R$, so the cathode-34 potential drops slightly; this increases the tube-$2B$ current, lowering point 38 so that the tube-$4B$ current decreases to zero.

12–9. Dead-zone Adjustment. When a motor moves a heavy roll to the right or left to seek a central position, as is described above in the side-register control, the motor-driven roll may coast beyond the center; before it stops moving to the left, the control circuit may give a signal to move it back to the right. To decrease this hunting action or back-and-forth movement of the roll, a dead-zone adjustment is included in Fig. $12I$, to let the roll overshoot the central position without causing an opposite correction. Tube 3 gives this dead-zone feature, as adjusted by $6R$.

If the $6R$ slider is turned away from $7R$ until the point-23 potential is, say, 200 volts above point 20, tube 3 will let enough current flow through $14R$ (electrons from 20 through $14R$ and tube 3 to positive point 25) to raise point 39 to about 205 volts above 20. Since 39 is also the cathode of tubes $4A$ and $4B$, both of these tubes may be passing a small amount of current (but not quite enough to pick up $1CR$ or $2CR$) inasmuch as points 37 and 38 are perhaps 200 volts above 20. With this setting of $6R$, a very small movement of the strip edge or small change of light will cause either the tube $4A$ or $4B$ current to increase slightly and pick up $1CR$ or $2CR$; such close adjustment may be expected to cause hunting or continued back-and-forth motion of the roll.

However, if $6R$ is turned clockwise, raising by 10 volts the potential at point 23, the tube-3 current increases, raising the voltage across $14R$ by nearly 10 volts.* Since cathode 39 now is perhaps 215 volts above 20, a greater change of light is needed at phototube 1 in order to raise point 37 or 38 by an extra 10 or 12 volts before relay $1CR$ or $2CR$ can cause a correction. In this way a dead zone is formed; when the wound roll approaches the central position and both $1CR$ and $2CR$ are dropped out, the roll, while coming to rest, may overshoot by the amount of this dead zone. Although the light changes at phototube 1, the voltage change at point 37 or 38 is not yet great enough to cause another correction. Of course, the wider dead zone permits less smoothness of the wound roll.

---

* Here tube 3 acts as a cathode follower. See Sec. 13–5.

CHAPTER 13

SPEED CONTROL OF D-C MOTORS

Most of the electronic controls of later chapters are used with d-c motors so as to produce variable-speed operation. These motors receive their direct current from various sources such as motor-generator sets, or from a-c supply lines through electron tubes.

13–1. The D-c Supply. In many cases the motor is supplied from a d-c bus or shop feeder whose voltage remains nearly constant. To raise the running speed of such a d-c shunt motor, the

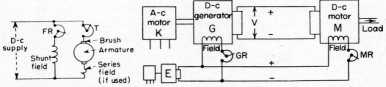

FIG. 13A. Connections and parts of a d-c shunt motor.   FIG. 13B. A variable-voltage d-c motor drive.

voltage across the shunt field must be decreased; a field rheostat may be used, such as FR in Fig. 13A, or the field may be controlled by thyratron tubes, as is described in Sec. 13–5. The motor speed may be reduced by inserting resistance in the armature circuit at T, but now a change in motor load also causes a change in speed; this is less desirable.

Often the d-c motor receives its armature current from a separate d-c generator, as shown in Fig. 13B. Although this generator G is driven at constant speed, its output voltage may be varied by control of the generator field; turning field rheostat GR clockwise increases the generator-field current, raising the output voltage V that generator G supplies to the armature of the d-c motor M, thereby increasing the motor-M speed. At the same time, motor M receives its field current from the smaller generator E; if the motor-field rheostat MR is turned clockwise, decreasing the motor-field current, the motor-M speed rises. This then is a variable-voltage d-c drive, wherein the motor speed may be varied by a change in either the motor-armature voltage or the motor-field voltage.

13–2. Armature Control and Field Control of Motor Speed. The speed of a d-c motor is controlled through a wide range by changing both the armature voltage and the field voltage as mentioned above. When the armature is connected across the largest voltage that the motor can stand continuously (called its rated voltage), and the field current is also at its largest or rated amount, the fully loaded motor runs at a medium speed, called its base speed. Figure 13C shows the usual way of operating a d-c motor below or above its base speed; in this example, the base speed is 1750 rpm. At point A the armature voltage and the field voltage are both at their largest amounts (250 volts), and this is at base speed. To reduce the speed, the field remains unchanged but the armature voltage is reduced (as by lowering the voltage supplied by the d-c generator). To make the motor run faster than base speed, the armature voltage is kept at 250 volts, but the field voltage (and current) is reduced by turning the field rheostat to increase its resistance. If the field current is decreased too far, the motor speed rises above a safe amount (as above point B in Fig. 13C).

The d-c motor usually is started with full field voltage, but with reduced armature voltage, so as to decrease the inrush or starting current.

To reverse the direction in which a d-c motor turns, the connections to either the field or the armature may be reversed (or, in Fig. 13B, the voltage produced by generator G may be reversed).

13–3. Tube Control of a D-c Motor. When a motor receives its d-c power from a variable-voltage generator as shown in Fig. 13B, small tubes may be added to control the field circuits of generator G and motor M, or the field of an exciter such as E. Such equipments are described in Chap. 15.

Instead of operating from a d-c generator, a variable-speed motor may receive power from an a-c feeder by using thyratron or ignitron tubes; these are arranged so as to change the alternating voltage into a direct voltage suitable for the d-c motor. Figure 13D shows such a thyratron arrangement; tubes A and B rectify the output of transformer T so as to apply a pulsing voltage to the armature, while tube C supplies a similar voltage to the motor field. Tubes

SPEED CONTROL OF D-C MOTORS

A and *B* may be controlled by their grids (and by a phase-shifting network, as described in Sec. 13-6 and later chapters) so as to vary the amount of voltage applied to the motor armature and thereby to change its speed. Such grid control also may provide smoother starting.

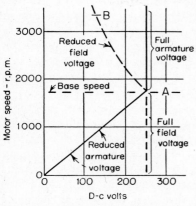

Fig. 13C. D-c motor operation above and below base speed.

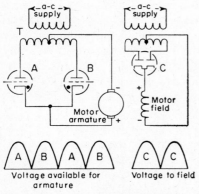

Fig. 13D. D-c motor operation from tube rectifiers.

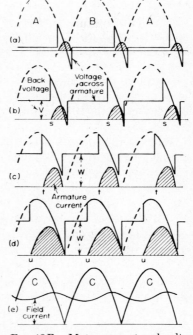

Fig. 13E. Motor current and voltage in Fig. 13D, varied by firing of thyratrons.

In (a) of Fig. 13E, the grid control delays the firing of tubes *A* and *B* until late in each half cycle, at points *r*. Current flows only during the small shaded portions, so the motor remains at low speed. To increase the motor speed, the tubes are fired earlier in each half cycle, at points *s*, as shown in (b). Notice that the motor now produces a d-c voltage of its own (shown at *V* and *W* in Fig. 13E, and called the motor's back voltage or *counter emf*); the height of this back voltage is an indication of motor speed.

To run the motor at full speed (at base speed, with full field current), the tubes are fired earlier, at t; if the load increases at the motor shaft, the tubes must be fired earlier, at u in (d), and current flows through the tubes and armature for a larger portion of each half cycle.

Meanwhile, the motor operates with full field current; as shown in (e) of Fig. 13E, this current is nearly constant during each cycle, for the large inductance of the field winding prevents much change in current.

Although in Fig. 13D only two tubes A and B are used to supply the armature current, larger motors may require so much current that three, four or six thyratron tubes are needed in this rectifier circuit. For motors larger than 50 hp, six ignitron tubes may furnish the armature current. In contrast, very small d-c motors may receive armature current from a single thyratron tube that supplies only a half-wave pulse during each a-c cycle, as described in Sec. 15–1.

When a medium-size d-c motor, receiving its armature current through thyratron tubes, must reverse its direction of turning, a reversing magnetic contactor often is used for this purpose instead of a second set of tubes. However, a small motor may receive half-wave armature current to make it turn clockwise, but may be driven in the reverse direction by half-wave armature current of opposite polarity, supplied through a second duplicate tube. This reversing arrangement is shown in Fig. 14E and described in Sec. 14–6.

In Fig. 13D only the armature current is changed by tube control, to vary the speed. When the motor must operate through a range of speed greater than 20 to 1 (so that the top speed must be more than 20 times as high as the bottom speed), both the armature current and the field current are varied by grid control of thyratron tubes (see Secs. 15–4 and 15–6). A smaller range of speed (such as 4 to 1) may be obtained by tube control of only the motor-field current, while the motor receives its armature current from a constant-voltage supply. As an example of this kind of drive, next let us study a motor whose speed responds to a beam of light.

13–4. Photoelectric Loop Control (CR7505–T102). Strip material such as steel or rubber is stored or supplied in large rolls; it must be unwound or fed off from such a roll, as is shown at A in Fig. 13F, before the strip passes into a mill operation at C.

SPEED CONTROL OF D-C MOTORS

Roll A is driven by a d-c motor M; the speed of this motor changes so as to keep a loop of the strip hanging between rolls A and B, thereby causing more constant tension in the strip entering at C. The loop hangs so as to cut off part of a large beam of light that is aimed to fall on a group of phototubes; the amount of light reaching the phototubes indicates the position of the loop. If the speed of motor M is too low so that roll A does not pay off the strip as fast as it is used at C, the loop rises, cutting off less of the light. As the light increases at the phototubes, the control panel weakens the field of motor M to increase its speed as needed.

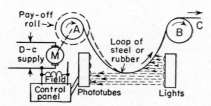

FIG. 13F. Arrangement of a photoelectric loop control.

When a new roll A is large, the motor may turn more slowly to feed the needed strip into the loop; as roll A becomes smaller, motor M and the roll must turn faster to feed the strip at the same speed.

The circuit of this loop control is shown in Fig. 13G. At the lower right, the armature and field of motor M are connected to a constant-voltage d-c supply; the field also may receive current from thyratron tubes 3 and 4. When these tubes pass no current, the top speed of motor M is set by rheostat R; the motor field current never is less than the steady flow from the d-c supply and through R. Whatever current is supplied through tubes 3 and 4 will add to the field current received through R; as tubes 3 and 4 are turned on, the total field current increases and motor M slows down.

If the steel or rubber loop reaches too low, it decreases the amount of light reaching phototubes 8 (center of Fig. 13G). As is explained below, this decreases the current in vacuum tubes 7 and 1, increasing the flow in tube 6A and saturable reactor 1SX; through a phase-shifting network (at lower left), transformer 2T is made to turn on tubes 3 and 4 so as to increase the motor field current, slow the motor and raise the position of the loop.

Before watching this circuit action in detail, notice that tube 2 rectifies the a.c. supplied through transformer $3T$; the pulsing voltage 48–5 is filtered by $1C$, $1X$ and $2C$ to produce 240 volts d.c. between top point 10 and bottom point 5. As is explained in Sec. 2–4, footnote, the voltage-regulator tube 5 and resistor $8R$ are added so as to hold steady the 150 volts d-c between points 9 and 5; jumper VR (shown above tube 2) is between two pins of VR tube 5, to remove voltage from the 150-volt bus if tube 5 is removed. A voltage divider or "ladder" (consisting of $3R$, $5R$, $6R$ and $7R$) holds point 3 at 80 volts; this sets the grid potential of tube $6B$. The voltage from 3 to the slider of $13R$ is applied across phototubes 8 and resistors $41R$ and $1R$. When no light reaches any phototube, there is so little current flowing through $1R$ that grid 14 of tube 7 is only a little more positive than point 13.

13–5. Cathode-follower Action. Tube 7 in Fig. $13G$ is used in a cathode-follower circuit; its load resistor $40R$ is below the tube, in the cathode circuit, instead of in the more usual location above the tube. The dashed line shows that this tube 7 is enclosed together with the phototubes 8 shown in Fig. $13F$. The phototube circuit has such high resistance (megohms) that its output signal across $1R$ is too weak to be sent by cable to the control panel. The cathode-follower tube-7 circuit does not increase the voltage of this signal, but it produces an equal signal across the lower resistance of $40R$ that may be carried safely by cable, to appear across $14R$ in the control panel. If grid 14 rises, cathode 1 follows or rises almost the same amount.

With no light on phototubes 8, grid 14 is slightly above point 13; tube 7 passes a little current through $40R$ so that the voltage across $40R$ is perhaps 4 volts greater than the voltage across $1R$. (Here the tube-7 grid voltage, measured between grid 14 and cathode 1, is -4 volts.) As the loop rises so that light reaches the bottom phototube, more current flows through $1R$ and grid 14 rises, say, 3 volts. This increases the tube-7 current through $40R$ so that cathode 1 rises about $2\frac{1}{2}$ volts (and the tube-7 grid voltage now is $-3\frac{1}{2}$ volts). So, as more light reaches phototubes 8 and grid 14 rises, the potential of cathode 1 follows in the same direction and by nearly the same amount. This voltage across $40R$ appears across the sensitivity adjuster $14R$, whose slider selects a part of this voltage signal for the control grid of pentode tube 1.*

---

* Adjuster $13R$ sets the tube-1 grid bias (which is the steady grid voltage to which the signal from 14 is added); $13R$ and $15R$ are adjustments to make

SPEED CONTROL OF D-C MOTORS

Tube 1, with its load resistors $9R$, $11R$ and $4R$, receives the steady 138 volts d.c. between points 9 and 41. Since $4R$ (added for stability) has very low resistance, the tube-1 cathode remains very close to point 41. As the loop rises (in Fig. 13F), increased light on phototubes 8 increases the tube-7 current and raises the control-grid potential of tube 1 in Fig. 13G; the tube-1 current increases and greater voltage appears across $9R$ and $11R$. Since the potential of tube-1 anode 25 drops, a similar but smaller drop occurs at the $15R$ slider; through $17R$, $20R$ and $21R$ this signal reaches the grid of tube 6A, to decrease the tube-6A current flowing through the d-c winding of saturable reactor $1SX$.* As is shown in Sec. 13–6, when the $1SX$ current decreases, the field current of motor M decreases, raising the motor speed.

Tubes 6A and 6B work as a long-tailed pair[12-8] with resistors $16R$ and $31R$, so the current increases in tube 6B as the tube-6A current decreases. By interchanging anode connections at X, the $1SX$ d-c winding may receive current from tube 6B instead of 6A so that, for other motor-control schemes, the increase in phototube light can cause an increase instead of a decrease in field current.

Notice that $16R$ can limit the current supplied to $1SX$, thereby limiting the maximum field current. Any current in $1SX$ must pass also through the resistance of $16R$; moreover, if the $16R$ resistance is increased, the current in $1SX$ and $16R$ causes cathode 56 to rise to higher potential, tending to turn off tube 6A by its own grid action.

13–6. Phase Shifting by Saturable Reactor. At the lower left in the loop-control circuit Fig. 13G, the a-c winding of saturable reactor $1SX$ is connected with $23R$ across the 50-volt winding of $S1T$, so as to shift the phase of the a-c voltage† applied to trans-

---

the full range of motor speed useful for controlling the loop. For operation as a pentode, the screen grid of tube 1 is at the steady positive potential of point 3, and the suppressor is connected to cathode.

* The portion within the dashed enclosure, between tubes 1 and 6A in Fig. 13G, is mostly for antihunt or stability purposes and is discussed in Sec. 13–7.

† This statement could imply that a sine wave of voltage is applied across $P2T$ and that the phase of this voltage is shifted. However, the operation of a saturable reactor is much more complicated than is indicated here, but the results of such operation can be described in this way, or as a shifting of the time of firing of the thyratron tube (within the a-c half cycle of its anode voltage).

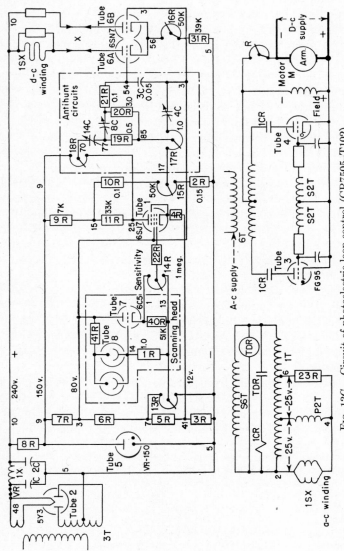

Fig. 13G. Circuit of photoelectric loop control (CR7505–T102).

former $P2T$, as is explained below. Two secondary windings $S2T$ are shown in the grid circuits of thyratron tubes 3 and 4.*
These phase-shifting circuits are shown again in Fig. 13H.

A saturable reactor such as $1SX$ is a variable inductance;† when no current flows in the direct-current winding (shown as $1SX$ above tube $6A$ in Fig. 13G), this reactor has large inductance or choke effect, which prevents the flow of much alternating current through its a-c winding. In a 60-cycle circuit, this reactor now permits the same amount of current to pass through it as would flow through a 30,000-ohm resistor. However, when about 1 or 2 ma of direct current flows through the many turns of the d-c winding, the inductance of SX has decreased so that it passes as much 60-cycle current as a 1000-ohm resistor.

If less than 1 ma flows in the d-c winding, so that the a-c winding of SX has 5000 ohms, the result is shown at (b) in Fig. 13H. Since SX and $23R$ each has 5000 ohms, the vector triangle‡ shows that the voltage (arrow G) applied to $P2T$ lags 90 deg behind the $S1T$ voltage; therefore, $S2T$ winding 1 applies to the tube-3 grid a wave of a-c voltage that lags 90 deg behind the tube-3 anode voltage (that is, the $S2T$–1 wave does not rise above the 0 line until a distance A later than the tube-3 anode rises above the 0 line). A half cycle later, the other $S2T$ winding 2 applies to the tube-4 grid a wave of a-c voltage that lags 90 deg behind the tube-4 anode voltage. The firing of each tube is delayed by an amount A; only half of each wave of a-c voltage (shaded) is used to force current through the field winding or load.

When the direct current in SX is increased to about 1.5 ma, the a-c winding has only 1200 ohms impedance. As shown at (c) in Fig. 13H, the voltage applied to $P2T$ lags by only the amount B, so the $S2T$ voltages fire tubes 3 and 4 quite early in their half cycles of anode voltage; nearly the entire voltage wave is used to force current through the load, and so this current is greater than before. However, when the direct current in SX is decreased so that the a-c winding has about 20,000 ohms, as is shown at (d),

---

* To make sure that thyratrons 3 and 4 carry no load current until their cathodes are fully heated, the $1CR$ contacts above these tubes do not close until the time-delay relay TDR has measured the required heating time.

† Chute, "Electronics in Industry," Chaps. 13 and 28.

‡ Since the voltage applied to $P2T$ generally is not a sine wave, this vector diagram can present only the *average* conditions within the cycle, or the resulting shift of the time of firing of the tubes.

the P2T and S2T voltages lag by the large amount C, firing tubes 3 and 4 very late in their half cycles of anode voltage; very little current is forced through the load.

13–7. Antihunt Circuits. Returning to Fig. 13G, we see that increased light on phototubes 8 (caused by a rising loop in Fig. 13F) will turn on tubes 7 and 1 and instantly will decrease the tube-6A

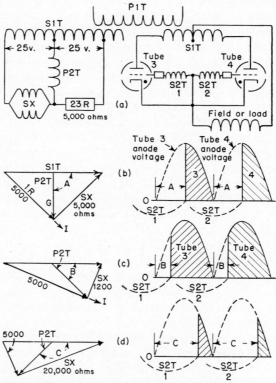

Fig. 13H. Phase-shifting by a saturable reactor, to vary a motor field and speed.

current through 1SX, thereby delaying the firing of thyratrons 3 and 4 (as indicated at d in Fig. 13H), decreasing the current applied to the field of motor M. The motor speed may increase so quickly that it lowers the loop too much, causing opposite action on all the tubes so that the motor speed decreases again. This up-and-down loop action or hunting may be decreased by the circuits within the dashed lines, between tubes 1 and 6A in

SPEED CONTROL OF D-C MOTORS 201

Fig. 13G. Since there is a considerable time lag between the light signal and the loop correction (we say that the system has large time constants), several antihunt circuits are provided, with separate adjusters 17R, 18R and tap switches (not shown) for selecting the size of the capacitors shown as 4C, 8C or 14C. These antihunt circuits are discussed further in Chap. 18.

When the loop rises and the increased light on phototubes 8 turns on tubes 7 and 1, anode 25 drops as is described in Sec. 13–4; there is a corresponding drop of potential at slider 17 of 15R. This is shown by the line AB in Fig. 13I; this straight-line control of motor speed may permit hunting. A small amount of "slow-down" action is provided by capacitor 3C, which must discharge through 20R and 21R; greater slowdown is produced by 4C.

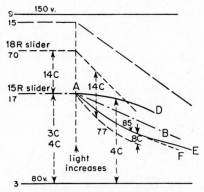

FIG. 13I. Antihunt action in circuit of Fig. 13G.

Before the light increases, Fig. 13I shows that 4C is charged to the voltage between 15R slider 17 and point 3; this also is the voltage between the grid of tube 6A and the tube-6B grid. When light increases, the charge on 4C prevents the tube-6A grid from following the line AB; instead, the grid potential follows the line AD, whose curved shape may be changed by varying the amount of capacitor 4C that must discharge through 17R. In this way the motor speed is made to change more smoothly.

Often a desirable antihunt action is obtained by forcing—that is, by causing a sudden overcorrection for a very short time, followed by very little correction until the resulting change in loop position occurs. Such overcorrection or "speed-up" action is provided by 14C and adjuster 18R. With steady light, Fig. 13I shows 14C charged to the voltage between 18R slider 70 and 15R slider 17; there is no voltage across 19R, so 8C has no charge. When the light increases, 18R may be set so that its slider-70 potential drops much faster than the 15R-slider-17 potential (line AB). Since 14C does not change its charge instantly, its terminal 77 drops as quickly as does slider 70; forcing downward point 85 and the tube-6A grid (as shown by line AE) to cause a quick change in motor speed. However, after 14C has charged to the new voltage condition (at the rate set by the size of 14C and the ohms of 19R and part of 17R), point 85 returns to the potential of 15R-slider 17. While 14C is thus changing its voltage, its charging current produces a voltage across 19R; capacitor 8C charges to this voltage. Later, as 8C discharges through 19R and 20R, the voltage thus produced across 20R further affects the grid potential of tube 6A (as indicated by line AF).

In Fig. 13I, the effect of 4C at line AD may seem to offset the effect of 14C at line AE. However, since the shapes of the slow-down and the speed-up waves can be adjusted separately, the final combined response wave may be quite different from the original line AB. These adjustments often are made by trial; approximate settings may be calculated, as is outlined in Chap. 18.

CHAPTER 14

PHOTOELECTRIC REGISTER CONTROLS

Motors are used to drive machines that cut printed paper strip at the proper spots so as to wrap packages or make bags correctly. The photoelectric equipments that make this possible are called cutoff or web-register controls;* several are described in Secs. 14–1 and 14–7. Similarly, other photoelectric equipments are called side-register controls; they cause motors to wind or split paper strip into rolls so that they are smooth on the sides. Such a control has been described in Sec. 12–7; others are explained in Secs. 14–4 and 14–5.

14–1. One-way Cutoff-register Control (CR7515–W201). This control equipment is shown in Fig. 14A; its diagram appears in Fig. 14B. This simple form of cutoff control acts only in one direction; unlike the CR7515–W108 control of Sec. 14–7, it cannot both decrease and increase the speed of the web. This continuous strip is fed from a large roll of paper; usually the main drive motor is geared to the draw rolls so as to feed this strip or web into the wrapping or bag-making machine slightly faster than the desired average speed. A spot has been printed at the edge of the strip, to permit this paper to be cut at the right point, between the patterns or pictures previously printed on the paper. As each printed spot passes beneath the scanning head (which contains the phototube and is connected by cable to the control panel shown in Fig. 14A), a signal is produced. Meanwhile, just as the cut is made, a cam on the cutter shaft operates a selector switch, between tubes 3 and 4. If the spot is "on time" or "in register," the selector-switch contact prevents the spot signal from reaching tube 4; there is no further action. However, when the spot arrives early, ahead of the cutoff knife, the spot signal occurs before the selector switch operates, so relay $1CR$ is picked up to retard or slow the web and keep the next spot in register.

We shall see that the spot signal received at phototube 1 causes

* Chute, "Electronics in Industry," Chap. 23.

a sudden dip in tube-3 current, turning on tube 5A; this decreases the tube-5B current and turns on tube 6 to pick up relay 1CR. After a time delay set by 2P, tube 5B again passes enough current to turn off tube 6 and drop out 1CR.

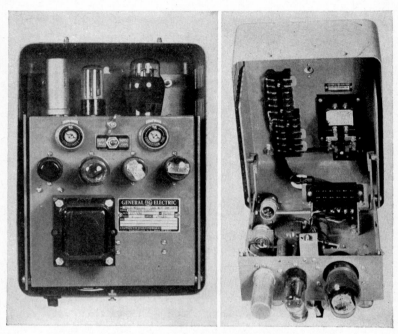

Fig. 14A. One-way cutoff-register control (CR7515–W201).

At the upper left in Fig. 14B, tube 7 rectifies the a-c control power supplied through transformer 1T, to produce 375 volts d.c. between upper point 10 and grounded point 5; this voltage is filtered or smoothed by capacitors 1C. Voltage-regulator tube 2 and resistor 1R serve to hold a constant 150 volts between points 22 and 5; as is described in Sec. 2–4, footnote, jumper VR is opened if tube 2 is removed. Part of the 1T voltage is rectified also by tube 4A (at lower center of Fig. 14B) so that bottom point 9 is at a d-c potential 165 volts below (more negative than) ground 5; this d-c voltage, filtered by 4C, serves only as a bias to keep tube 6 from firing.

To permit this equipment to respond to a printed spot that is either lighter or darker than the web background, switch 1S is included at phototube 1 in Fig. 14B.* For operation on light increase, 1S shorts 5R; light causes more electrons to flow from point 5 through 1S, phototube 1 and 4R to the +75-volt potential at point 14. The voltage increases across 4R, driving points 16 and 24 and tube-3 grid 26 more negative; this causes relay 1CR to pick up, as is explained later. However, when 1S is in the position for operation on light decrease, 4R is shorted and the phototube current produces voltage across 5R. The decrease of phototube current and 5R voltage drives points 15 and 24 and grid 26 more negative, causing relay 1CR to pick up.

14–2. Response Only to Sudden Light Changes. In Fig. 14B, when a change of light at phototube 1 reduces the voltage at point 24 as described above, tube 3 and relay 1CR will operate only if the light changes quickly, as when a printed spot enters the light beam suddenly. This kind of action is caused by "capacitor coupling"; phototube 1 is coupled to tube 3 through capacitor 2C.

When a constant or steady light reaches phototube 1, there is a steady voltage drop across 4R or 5R (depending on the setting of switch 1S) and there is steady voltage between point 24 and ground 5; capacitor 2C is charged to this same voltage. Current flows through 1P only while 2C is charging or discharging. While steady voltage remains between 24 and 5, there is no voltage drop across 1P, so the tube-3 grid 26 is at the same potential as point 5.

If the amount of light on phototube 1 now slowly decreases (with 1S set upward for operation on light decrease), the current decreases through phototube 1 and 5R. While the voltage across 5R decreases slowly, 2C discharges to this reduced voltage by forcing a very small current to flow through 1P; the tube-3 grid is not affected much. No matter how much or how little light shines on phototube 1, capacitor 2C has charged to the voltage across 1P, and grid 26 is still at the same potential as point 5. Meanwhile cathode 17 of tube 3 is 1 volt above 5, so the grid voltage is -1 volt; tube 3 passes enough current to hold its anode 27 steadily at about +70 volts (70 volts more positive than 5).

When a printed spot suddenly reduces the amount of light reaching phototube 1 (and the selector switch has not yet operated), the current through 5R decreases sharply, and reduces the voltage

* Phototube 1 is mounted in a scanning head, separate from the relay enclosure. Different types of phototube and scanning head may be used for response to reflected or transmitted light.

206 ELECTRONIC MOTOR AND WELDER CONTROLS

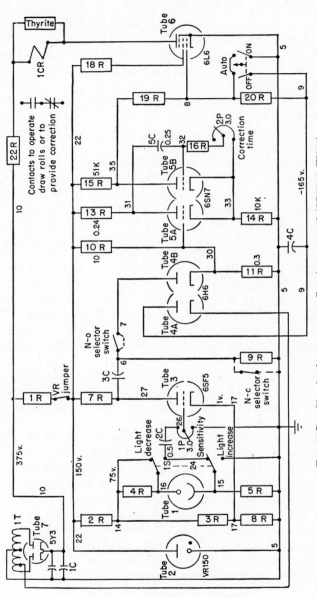

Fig. 14B. Circuit of one-way cutoff-register control (CR7515–W201).

drop across $5R$. Capacitor $2C$ cannot discharge through the resistance of $1P$ fast enough to follow this abrupt change of voltage across $5R$; this discharge current causes a voltage to appear across $1P$ so that grid 26 is forced more negative than point 5.* The tube-3 current decreases quickly and anode 27 rises, as mentioned next. If the printed spot stops (or a large spot moves slowly past), still reducing the light at phototube 1, capacitor $2C$ discharges within several seconds to the new low voltage across $5R$, and anode 27 returns at this same rate to its $+70$-volt potential.

We shall see that tube 3 also is capacitor-coupled to tube 5 through capacitor $3C$. Before the tube-3 current changes, $3C$ has charged to the steady voltage between anode 27 and 5; there is no current through $9R$, so terminal 6 of $3C$ is at the same potential as ground 5. Although the n-o (normally-open) selector switch connects points 6 and 7, no current flows through tube $4B$; the tube-$4B$ cathode 30 is held about 4 volts above point 5, by the voltage divider $10R$ and $11R$. When the printed spot affects phototube 1 so that the tube-3 anode 27 rises (as mentioned above), the voltage across $3C$ cannot change quickly, so its terminal 6 rises also.† If the n-o selector switch has been closed by the machine, anode 7 of tube $4B$ is driven higher than the cathode-30 potential. Since tube $4B$ is a simple diode rectifier, the rise of anode-7 potential lets electrons flow from ground 5 through $11R$ and tube $4B$ to help charge $3C$; this added electron flow increases the voltage drop across $11R$, so that the potential of point 30 rises sharply. Since point 30 is also the grid of tube $5A$, we see that the tube-$5A$ current increases suddenly (to turn on tube 6 as is described later) whenever the tube-3 anode 27 rises sharply. However, a slow rise at 27 produces too little effect at tube $5B$, since $3C$ can be charged through $9R$ in about $\frac{1}{4}$ sec. Also, if the anode-27 potential drops suddenly (as occurs when light returns to phototube 1 after a large printed spot has passed from its view), the voltage across $3C$ forces its terminal 6 more negative

---

* The position of the $1P$ slider determines how much of this signal voltage shall reach the tube-3 grid 26; if turned closer to 5, there must be greater light change at phototube 1 before relay $1CR$ is picked up, so the sensitivity of the control is decreased.

† If an n-c (normally-closed) selector switch is used in Fig. 14B, point 6 cannot rise until the selector switch has opened. A jumper 6-to-7 permits the point-6 potential to reach the anode of tube $4B$.

than point 30; tube $4B$ prevents such negative signals from reaching tube $5A$, for these could disturb the correction-time circuit of tube $5B$, next to be described.

14–3. Correction-time Circuit. Until a signal is received through tube $4B$ in Fig. $14B$, grid 30 of tube $5A$ is about $4\frac{1}{2}$ volts above point 5. During any such steady condition, capacitor $5C$ has taken a steady charge and draws no current through $16R$ or $2P$, therefore there is no voltage across these resistors. Since $5C$ terminal 32 (connected to the grid of tube $5B$) is at the same potential as cathode 33, tube $5B$ has zero grid voltage and permits a current of nearly 2 ma to flow through tube $5B$ and resistors $14R$ and $15R$. This current produces about 18 volts drop across $14R$, so cathode 33 of tubes $5A$ and $5B$ is about 18 volts above 5. Therefore tube $5A$ has about -14 grid voltage and no current flows through $13R$ or tube $5A$; its anode 31 is near the high potential of point 22, so $5C$ is charged to the steady d-c voltage between points 22 and 33. Because of the current through $15R$ and tube $5B$, anode 35 is about 40 volts above 5; between 35 and the -165-volt potential of bottom point 9, the voltage divider $19R$–$20R$ holds grid 8 of tube 6 at about 45 volts below cathode 5, so tube 6 passes no current and relay $1CR$ is not picked up.

When a printed spot makes phototube 1 send a signal through tubes 3 and $4B$ to raise the potential of grid 30, current flows through tube $5A$ and $13R$, quickly lowering the potential of terminal 31 of $5C$. Because the voltage across capacitor $5C$ cannot change quickly, its terminal 32 and the tube-$5B$ grid are driven far negative, stopping the flow of current through $15R$ and tube $5B$; anode 35 rises toward point 22, and grid 8 of tube 6 also rises, letting tube 6 pass current to pick up relay $1CR$.

Although this "turn-on" positive signal at grid 30 may last only $\frac{1}{1000}$ to $\frac{1}{8}$ sec, notice how the tube-$5A$ current is made to continue for a longer time, such as may be needed for the tube-6 current to pick up relay $1CR$. The tube-$5A$ anode resistor $13R$ is selected to have nearly five times as much resistance as has the tube-$5B$ anode resistor $15R$; therefore the greatest current that may flow through tube $5A$ is much less than the greatest tube-$5B$ current. When the signal turns on tube $5A$, which turns off tube $5B$, the current flowing through cathode resistor $14R$ decreases from 2 ma to about 0.4 ma, so the cathode-33 potential decreases until it is only 4 volts above point 5. When the brief signal has passed and grid 30 returns to its steady value ($4\frac{1}{2}$ volts above 5), the grid-to-cathode voltage of tube $5A$ is now $+\frac{1}{2}$ volt, so the tube-$5A$ current continues to flow, thereby holding its anode 31 at a potential far below that of point 22.

Electrons now flow from $5C$ terminal 32, through $16R$ and $2P$ to 33, discharging $5C$. After a time delay as set by the resistance of correction-time adjuster $2P$, the voltage across $5C$ has decreased enough to let the tube-$5B$ grid 32 return close to the cathode-33 potential. Tube $5B$ again passes current, lowering its anode 35 and the tube-6 grid 8 so that the tube-6 current stops, dropping out relay $1CR$. As current returns to tube $5B$, it increases the voltage drop across $14R$ so that cathode 33 rises and tube $5A$ is turned off, to await the next signal from phototube 1.

The length of this correction time is adjusted by $2P$ so that relay $1CR$ is picked up for the desired time; its contacts may operate some feed-correcting motor for a time just long enough to correct for the error in cutoff operation.

For automatic operation of this control, both contacts of the off-on switch are open, at the lower right in Fig. $14B$. To pick up relay $1CR$ steadily, the switch is turned to connect grid 8 of tube 6 to point 5; this gives zero grid voltage, so tube 6 passes current. To prevent relay $1CR$ from being picked up at any time, the switch is turned to connect grid 8 to point 9; this -165-volt grid prevents any flow of tube-6 current.

14–4. Side-register Control (CR7515–S127). While a web-register control "sees" printed marks pass lengthwise at high speed, a side-register control "looks" at the edge of such a web (of paper, rubber, metal, etc.) and "sees" only a small sidewise movement while the web is being wound onto a roll. Figure $14C$ shows such a control; its diagram appears in Fig. $14D$, and includes an amplidyne generator to supply voltage for the correction motor. When the web is lighter than the background, phototube 1 receives more light if the web shifts toward one side, but it receives less light if the web shifts the other way. To keep the edge of the web from moving more than, say, $\frac{1}{64}$ inch sidewise, the circuit of Fig. $14D$ must respond to small and slow changes of light as the web shifts. No tubes in this circuit are capacitor-coupled.[14–2] Many refinements are added to protect this circuit from outside voltage dips. The d-c supply from rectifier tube 8 and its filter ($1C$, X, $2C$) is further smoothed by voltage-regulator tube 3, so that the voltage 26-to-3 is held constant at 150 volts (as is explained in the footnote to Sec. 2–4). Tube 2 is a pentode, whose filament current is held constant by ballast tube 7.* Amplifier tube 2 is mounted with phototube 1 in the scanning head, remote from the control panel, to increase the strength of the signal before it passes through the cable to the rest of the circuit.

Fig. 14C. Side-register control (CR7515–S127).

When phototube 1 sees the web shifting to one side, a d-c correction motor (upper right in Fig. 14D) moves the entire roll sidewise, to bring the web edge back to the right place; this makes a smooth-edged roll. This d-c motor has constant field current (from another rectifier tube, not shown); the motor armature voltage comes from an amplidyne generator† (driven at constant speed

* A ballast tube is not electronic, for its current flows in a filament, like the current in an incandescent lamp. Its filament (usually made of iron) is heated by the load current passing through it. Within its operating range, any increase of load current raises the temperature of part of its filament and increases the resistance of the filament. After a short time lag, this resistance has increased the right amount to bring the current back to its previous value.

† Chute, "Electronics in Industry," Chap. 28. The amplidyne generator acts like an ordinary d-c generator, with the added ability to produce quickly a large change in output power in response to a very small current in its control fields. Thus the amplidyne acts as an amplifier; in Fig. 14D, tubes 5

by an a-c motor). Beam power tubes 5 and 6 control the d-c fields of this generator so as to change the amount of voltage supplied to the correction-motor armature; in this way the d-c motor is turned in either direction or is stopped.

Some one position of the web is "just right"; at this position the d-c correction motor must not turn. At this best position, phototube 1 receives a certain amount of light and sets the control-grid-10 voltage of tube 2; however, the amount of tube-2 anode current can be increased by turning $10R$ clockwise (lowering the

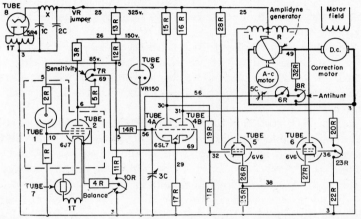

FIG. 14D. Circuit of side-register control (CR7515–S127) with amplidyne generator.

cathode potential of tube 2). At the right setting of $10R$ we shall see that tube 5 passes the same amount of current* as tube 6 passes; the current in amplidyne field L is equal to the current in field R. These fields "buck," or are opposed, so the generator sends no voltage to the d-c motor; the motor does not turn.

Let us see how phototube 1 controls tubes 5 and 6 and the d-c motor in Fig. 14D. When the web shifts and increases the light on tube 1, grid 10 rises and tube 2 passes more current; more electrons flow from 3 through $10R$, $4R$, tube 2 and $3R$ to the constant point 26, also through $5R$ and $7R$ to the constant point 5. Greater tube-2 current lowers anode 6 and the potential of the $7R$ slider,

---

or 6 pass perhaps 20 ma to the control fields L and R, at once causing the amplidyne to furnish 20 amp to drive the correction motor.

* The anode current of tubes 5 and 6 may be balanced further by turning $23R$.

connected to the grid of tube $4B$;* this decreases the current in tube $4B$ and $16R$. Notice that tubes $4A$ and $4B$ are a "long-tailed pair";[12-8] less current in $4B$ lowers cathode 29, increasing the current in tube $4A$, whose grid is at the constant potential of point 5. In this way, the lowering of $4B$ grid 69 raises the anode potential at 31, but lowers the anode-30 potential. The potential rises at $23R$ and grid 36, turning on tube 6 and strengthening the R field; the grid-32 potential falls, turning off tube 5 and weakening the L field. Since field R is now so much stronger than the opposing field L, the generator sends a mature voltage to turn the correction motor, which moves the roll so as to decrease the light reaching phototube 1.

When less light reaches tube 1, turning off tube 2, this turns on tube 5 and turns off tube 6; the L field now is stronger than the R field, reversing the amplidyne output voltage so that the correction motor now moves the roll so as to increase the light reaching phototube 1.

In Fig. $14D$, grid 56 of tube $4A$ connects to the antihunt circuits including variable capacitors $3C$ and $5C$, also $14R$; the voltage between point 3 and the $8R$ slider is applied across $3C$ and $5C$ connected in series. So long as the amplidyne output voltage is steady or is changing slowly, the charge on $3C$ and $5C$ is nearly steady, so no current flows through $14R$; grid 56 is at the same potential as point 5. However, whenever the amplidyne voltage (49 to 3) changes quickly, the potential at junction 56 changes for a moment. This change may tend to turn tube $4A$ on or off, depending on the capacities selected for $3C$ and $5C$ by trial, to decrease the system hunting; the time length of this antihunt signal increases as more resistance is inserted in $6R$.

14–5. All-tube Side-register Control (CR7515–S120). Whereas the control described above uses an amplidyne generator to furnish d.c. to the correction motor, Figs. $14E$ and $14F$ show the circuit of a side-register control wherein thyratron tubes 8 and 9 supply

* If the $7R$ slider is moved toward $5R$, a larger portion of the point-6 signal will be used to control tube $4B$; this increases the sensitivity of the control, since a smaller sidewise roll movement now causes the correction motor to turn. This equipment includes a "sense-reversing" switch (not shown) whereby the signal at the $7R$ slider may be connected to the tube-$4A$ grid instead of to the tube-$4B$ grid. In this way the operator may select the sense of response, or the direction of roll movement corresponding to an increase of light at phototube 1. This switch also reconnects the antihunt circuits to correspond to the desired sense or direction of roll movement.

current direct to the correction motor; tube 8 fires to make the motor move the roll sidewise in one direction, but tube 9 causes roll movement in the opposite direction.

That part of the circuit shown in Fig. 14E is used also as part of a web-register control shown in Fig. 14H and described in Sec. 14–7, when combined with Fig. 14I. In Fig. 14E, a supply of 335 volts d.c. is provided by rectifier tube 1 and the filter (3C, X, 4C). Tubes 3A and 3B respond to signals received (at grid terminals 82 and 83, at upper right) from either the side-register circuit of Fig. 14F, or the web-register circuit of Fig. 14I. As will be explained later, when grid 83 is made more negative, less current flows through tube 3A and its cathode res'stor 58R. Since tubes 3A, 3B and 58R operate as a "long-tailed pair," the lowering of cathode 16 gives the same result as a rise of grid-82 potential; greater current flows in tubes 3B, 4B and the d-c winding of saturable reactor 1SR; no current flows in 2SR. Lower in Fig. 14E, greater 1SR current fires thyratron tube 8 earlier in its half cycle of a-c anode voltage, while tube 9 fires later or not at all. The current pulses from tube 8 (flowing always in the same direction) pass through the armature of a d-c correction motor, turning this motor so as to give the desired movement to keep the roll or web in register. Before studying Fig. 14E in detail, let us see how the side-register circuit in Fig. 14F produces a signal that lowers the grid-83 potential as outlined above.

In Fig. 14F the phototube 12, pentode 13, voltage-regulator tube 10 and ballast tube 11 act like tubes 1, 2, 3 and 7 in Fig. 14D, described in Sec. 14–4. When the web roll is in the correct position, no motion of the correction motor is desired. At this position, phototube 12 receives a certain amount of light and sets the control-grid-10 voltage of tube 13; with this steady grid voltage, the amount of tube-2 anode current can be decreased by turning 29R clockwise, raising the cathode potential of tube 13. At the right setting of 29R we shall see (in Fig. 14E and Sec. 14–6) that both tubes 3A and 3B pass very little current, so that thyratrons 8 and 9 are held off equally and the correction motor does not turn.

When the web shifts sidewise and increases the light on phototube 12, grid 10 rises and tube 2 passes more current; more electrons flow from 3 through 29R, tube 13 and 23R to the constant point 90, also through 31R to the constant point 5. As greater tube-13 current lowers anode 6, the potentials at slider 92 and

214   ELECTRONIC MOTOR AND WELDER CONTROLS

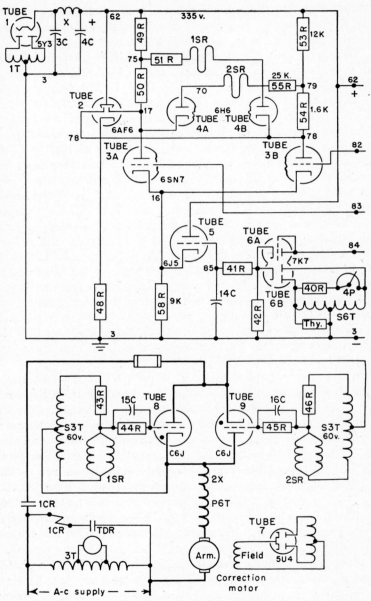

Fig. 14E.   Correction-motor circuit, controlled from Fig. 14F or Fig. 14I.

point 83 also are lowered. By turning 31R clockwise, a larger portion of the signal at 6 is used; the circuit becomes more sensitive, for the correction motor responds to a very small shift of the web. With such quick action the equipment may "hunt"—the motor moves the roll too far to one side and then too far in the other

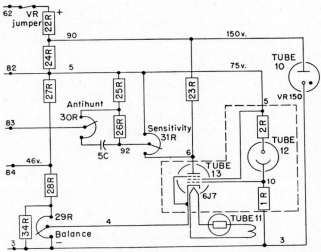

Fig. 14F. Circuit of all-tube side-register control (CR7515–S120) including Fig. 14E.

direction. To decrease such hunting, part of the voltage between anode 6 and point 5 (fixed by the voltage divider 24R, 27R, 28R and 29R) is applied across 5C and 30R. When light increases at phototube 12 and point 6 drops to a lower potential, the voltage across capacitor 5C does not change at once; grid 83 is forced negative (as though connected directly to anode 6) but it then returns more positive as 5C discharges through 30R. In this way, the d-c correction motor is forced to turn quickly at first, then it slows down or stops until the effect of its correction can be seen at phototube 12. Now let us see how the circuits in Fig. 14E convert this signal at grid 83 so as to control the speed and direction in which the motor turns.

14–6. Web Correction by Saturable Reactor. In Fig. 14E first let us study the steady circuit conditions before a correction signal is received at grid 82 or 83. To produce this balanced condition (that prevents motion of the correction motor), grids

82 and 83 both are at about the same potential, as finally adjusted by the balance potentiometer ($29R$ in Fig. $14F$ or $5P$ in Fig. $14I$).*
Tubes $3A$ and $3B$ are passing about 5 ma each; these electrons flow from grounded point 3 through $58R$ to cathode 16, through tube $3A$, $50R$ and $49R$ to 62, also equally through tube $3B$, $54R$ and $53R$ to 62. Cathode 16 of tubes $3A$, $3B$ and 5 is perhaps 80 volts above 3; the tube-5 grid 85 is 30 or 40 volts more negative than 16 (as is shown later) so no electrons flow through tube 5. Above tube $3A$, point 75 is about 7 volts more positive than anode 78 of tube $3B$, so part of the tube-$3B$ electrons flow through tube $4B$ and the d-c winding of saturable reactor $1SR$ and $51R$ to point 75, then through $49R$ to 62. Similarly, a part of the tube-$3A$ electrons flow through tube $4A$, $2SR$, $55R$ and $53R$ to 62. These small direct currents flowing equally through $1SR$ and $2SR$ make both thyratron tubes 8 and 9 fire late in their half cycles as is shown at (b) in Fig. $14G$ and explained below. Tubes 8 and 9 are connected back to back across the a-c supply; when each tube passes the same amount of current, a small a-c voltage appears across the motor armature; there is no d-c voltage to make the motor turn either way, although the motor receives full field current through rectifier tube 7.

To make the motor turn one way, the current through tube 8 must be made greater than the tube-9 current. The thyratron-8 current is gradually increased or decreased by the phase-shifting bridge circuit,† shown in Fig. $14G$. This circuit includes a constant resistor $43R$ and a variable inductance, which is the saturable reactor $1SR$. When no direct current flows in tube $3B$ or the d-c winding of $1SR$, this reactor has large inductance or choke effect, which prevents the flow of much alternating current; in a 60-cycle circuit, this reactor now permits the same amount of current to pass though it as would flow through a 30,000-ohm resistor. However, when about 1 or 2 ma of direct current flows through tube $3B$ and the many turns of the $1SR$ d-c winding, the inductance of $1SR$ has decreased so that it passes as much 60-cycle current as a 1000-ohm resistor.

If less than 1 ma of d.c. flows in $1SR$, so that it has 5000 ohms, the result is shown at (a) in Fig. $14G$. Since $1SR$ and $43R$ each

---

* In Fig. $14F$ grid 82 is held at a constant potential 75 volts above point 3. In Fig. $14I$, grid 82 is at the constant potential of point 8, 90 volts above point 3.

† Chute, "Electronics in Industry," Chap. 13.

has 5000 ohms, the vector triangle shows that the tube-8 grid voltage (arrow G) lags 90 deg behind the anode voltage (or behind the $S3T$ voltage, in phase with the anode voltage), as shown by A; tube 8 fires late, delayed by the amount A, and applies voltage to the motor armature for about half of each wave.*

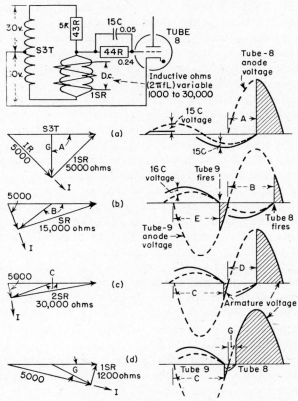

Fig. 14G. Phase-shifting by saturable reactor in Fig. 14E.

In the balanced condition mentioned earlier, where only small equal direct currents flow through either $1SR$ or $2SR$, this smaller d.c. in $1SR$ has increased the ohms of its a-c winding to perhaps

* To prevent accidental firing of thyratron tube 8 at the start of its half cycle, we insert $44R$ and $15C$ in the grid circuit, as shown in Fig. 14G. By grid rectification, $S3T$ forces current through $44R$, charging $15C$ so that it is more negative at the grid connection. The result is shown in (a); its effect is important when the a-c "holdoff" wave is nearly 180 deg out of phase with the anode voltage, like the condition shown for tube 9 in part (c).

15,000 as shown at (b). The tube-8 grid voltage lags by the amount B, so tube 8 fires very late, and applies voltage to the motor armature for a small part of the half cycle. While the circuit is in balance, $2SR$ also receives this small amount of d.c., so thyratron 9 also fires after a delay E behind its own anode voltage; in (b) tubes 8 and 9 pass the same amount of current, but in opposite directions, so the d-c motor does not turn.

Now see what happens when a correction signal is received at point 83 in Fig. 14E, so as to decrease the tube-3A current. Less current flows through 58R or th ough 49R and 50R; the potential rises at anode 17 and point 75, but drops at cathode 16 because of less voltage drop across 58R. Meanwhile grid 82 of tube 3B is held at steady potential (by the circuit of Fig. 14F or Fig. 14I); since the tube-3B cathode 16 is lowered, the current increases in tube 3B, lowering anode 78 and point 79. Notice that the correction signal at 83 raises 75 and lowers 78; this increases the voltage and current flow through 51R, 1SR and tube 4B. Also, since 79 is lowered and 17 rises, cathode 17 of tube 4A becomes more positive than its anode 70, so no d.c. flows in 2SR. Part (c) of Fig. 14G shows that, when there is no d.c. in 2SR, the firing of tube 9 is delayed by a greater amount C, until the current remaining through tube 9 is negligible. Meanwhile the increased d.c. in 1SR has caused earlier firing of tube 8, similar to that shown in part (a), until tube 8 is delayed only by the amount D. Since the positive voltage applied to the armature by tube 8 is greater than the negative voltage (below the line) applied by tube 9, the motor armature turns at medium speed (to retard the web or to move the roll, say, to the left).

If a stronger correction signal at 83 further decreases the tube-3A current, this cannot shut off tube 9 further; however, the tube-3B current increases, lowering anode 78 to permit a greater flow of d.c. in 1SR. Under this new condition, shown in (d) of Fig. 14G, tube 8 fires after a small delay G; much greater voltage is applied to the motor armature, which turns at high speed to cause faster correction.

To produce an opposite correction (such as is required when grid 83 is driven positive in Fig. 14F, or when grid 82 is pulled negative in Fig. 14I), the d.c. decreases in 1SR and increases in 2SR; tube 8 fires very late but tube 9 fires early, applying a large voltage of opposite polarity (below the line) to the armature so that the motor advances the web or moves the roll, say, to the right.

PHOTOELECTRIC REGISTER CONTROLS 219

FIG. 14H. Two-way cutoff web-register control (CR7515–W108).

Tube 2 in Fig. 14E is an electron-ray or indicator tube; its green-light circle or "magic eye" winks to show what web correction is being made. Electrons flow from grounded point 3 through 48R and from cathode to circle anode of tube 2 to positive point 62. These electrons may strike all parts of the tube anode circle, making a green glow over the entire circle. However, this tube has two deflecting wires, shown connected to points 17 and 78. When tubes 3A and 3B pass equal currents, anodes 17 and 78 are more negative than the tube-2 anode 62; each deflecting wire of tube 2 repels electrons away from a small part of the green circle anode, so this "eye" shows a small shadow wedge on each side of the circle. When a negative signal at 83 makes anode 17 rise and anode 78 fall, the shadow near wire 17 decreases, while the shadow near wire 78 becomes larger, indicating that thyratron 9 is applying greater voltage to turn the motor.

14–7. Two-way Cutoff Web-register Control (CR7515–W108).

The circuit in Fig. 14E, explained above, may be combined with the circuit in Fig. 14I to form the high-speed web-register control next described. The panel that mounts the parts in the Fig. 14I circuit appears in the upper part of Fig. 14H; the main or lower portion of Fig. 14H is diagramed in Fig. 14E.

As is shown at the lower right in Fig. 14I, the draw rolls pull the sheet or web off of a roll of material, previously printed with a design and register marks, accurately spaced. This web is fed between cutter rolls that try to cut the web in line with a register mark. The cutters are driven at constant speed, but the speed of the draw rolls is changed or controlled (by the correction motor in Fig. 14E) so as to feed the web forward just fast enough to bring a register mark at each cutoff point. A beam of light X is so located that a preceding register mark decreases the light reaching phototube 18, at the exact instant when a register mark reaches the cutoff point. If the register mark is early, so that this light dip at phototube 18 occurs before the cutoff knives have turned to the cutting position, the circuit of Fig. 14I makes thyratron tube 8 pass current (in Fig. 14E); the correction motor turns and (through a differential gear) slows the draw rolls to retard the web.

Like phototube 18 "sees" when the register mark reaches the cutoff point, phototubes 15 and 16 "see" when the knife makes its cut. In the selector-switch assembly (shown as a dashed enclosure in Fig. 14I) there is a turning disk, driven by chain or gear from the cutoff roll; through pairs of holes in this disk, beams of light

PHOTOELECTRIC REGISTER CONTROLS 221

may reach phototubes 15 and 16. For an instant before the cut is made, light beam Y reaches phototube 15, trying to turn on tube 11 and thyratron 8, as is later explained. Just after the cut is made, light Z reaches phototube 16, trying to turn on tube 12 and thyratron 9. By a mechanical "dead-zone" adjustment, the holes in the turning disk may be located so that no light reaches

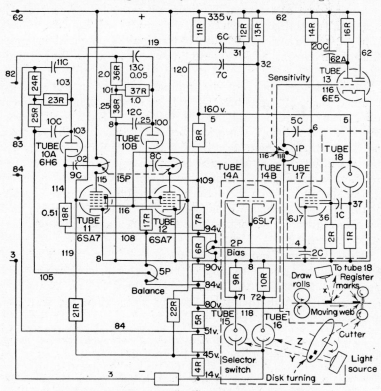

Fig. 14I. Circuit of web-register control (CR7515–W108).

either tube 15 or 16 at the time of cutting. If the register-mark light dip at phototube 18 occurs during this dead zone, then neither tube 11 nor tube 12 passes current and the correction motor does not turn; the web speed is "just right."

All tubes in Fig. 14I operate from the d-c supply received between terminals 62 and 3, from tube 1 (and filter 3C, X, 4C) in Fig. 14E. A voltage divider (4R to 11R, center of Fig. 14I) furnishes the various d-c voltages as shown. Briefly phototube 18 controls pentode 17 and the control grids or first grids of tubes 11 and 12; phototubes 15 and 16 control tubes 14A and 14B and the third grids of the "mixer" tubes 11 and 12. This type of tube passes current only when both its first and third grids are positive; other grids prevent any interaction between the signals received at the first and third grids. When tube 11 passes current, the potential dips at terminal 83, increasing the tube-8 current (in Fig. 14E, as is described in Sec. 14–6), to retard the web. When tube 12 passes current, the dip in terminal-82 potential increases the tube-9 current, to advance the web. Now let us study this in detail.

Phototube 18 and pentode amplifier tube 17 in Fig. 14I are coupled through capacitor 1C; only a sudden light change can affect tube 17, as is explained in Sec. 14–2. Each dark register mark on the web dips the grid voltage of tube 17 in the scanning head; its anode current decreases, so its anode 6 rises quickly and the potential rises at all parts of sensitivity adjuster 1P. This raises the control grids 116 of tubes 11, 12 and 13 (whose potential at the 80-volt point on the divider had been 10 volts more negative than the tube cathodes connected to point 8). Therefore, each passing register mark makes tube 13 "wink" and tries to turn on tubes 11 and 12.*

This "turn-on" impulse is shown at A in Fig. 14J; it will not turn on tubes 11 or 12 as long as the impulse occurs during "dead zone" B. During B, the third grids of tubes 11 and 12 are both so negative that neither tube can fire, even when the first grids are positive.

* Tube 13 is an electron-ray or indicator tube responding to low-level signals. It shows a circle of green light. The left-hand side of tube 13 (in Fig. 14I) is a triode amplifier. When grid 116 of this triode is too negative, no electrons flow from 8, cathode to anode 62A, or through 16R to point 62; points 62A and 62 are at the same potential, so electrons flow freely from cathode to right-hand anode 62 of tube 13. These electrons strike all parts of the anode circle, making a green glow over the entire circle. Notice the electrode or wire below circle anode 62, connected to anode 62A inside tube 13. When a signal voltage raises the potential of grid 116, current flows through 16R so that point 62A becomes more negative than point 62; capacitor 20C charges to this voltage across 16R. Since the electrode is more negative than anode 62, it repels some of the electrons away from the circle; a dark wedge, or shadow, appears in the green circle. When grid 116 rises for an instant, because of the passage of a register mark, the shadow wedge becomes larger, then gradually closes again as 20C discharges through 16R; the "magic eye" seems to wink.

The holes in the turning disk are adjusted so that no light reaches phototubes 15 or 16 during zone B. With no current through these phototubes, points 71 and 72 (grids of tube 14) are at cathode potential 8, so tube 14 passes current through both $12R$ and $13R$; anode potentials 31 and 32 are low. The third grids 119 and 120 are about 45 volts more negative than cathode 8.

Earlier than zone B, a disk hole lets light reach phototube 15; current through $9R$ lowers grid 71, turning off tube $14A$. Anode 31 rises, and $6C$ raises the third grid 119 also,* as shown during zone G of Fig. $14J$. Also, later than zone B, a different disk hole

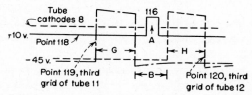

Fig. $14J$. Grid potentials in circuit of Fig. $14I$, with spot in register.

lets light reach phototube 16; current through $10R$ lowers grid 72, so anode 32 rises and the third grid 120 of tube 12 becomes positive, during zone H. Neither of these actions turns on tube 11 or 12, for their first grids are too negative to permit electrons to flow. As long as signal A from the register mark stays within zone B, tubes 11 and 12 give no correction signal.

So long as tubes 11 and 12 pass no current, the sliders of correction-rate adjuster $15P$ are at high potential near point 62; the cathodes of tubes $10A$ and $10B$ are at the lower potential of point 108, so capacitors $8C$ and $9C$ are charged to this 62-to-108 voltage. The anode of tube $10B$ is connected through $37R$ and $38R$ to the potential at point 8; since 8 is more negative than 108, rectifier tube $10B$ passes no current. Terminal 82, at point-8 potential, holds the tube-$3B$ grid in Fig. $14E$ about 90 volts above grounded point 3; similarly terminal 83, adjusted by the balance dial $5P$, holds the tube-$3A$ grid also about 90 volts above 3, so that the thyratron-8 current is equal to the thyratron-9 current and the correction motor does not turn.

14–8. Action to Retard or Advance the Web. Suppose that the web is moving too fast, so that the register mark produces signal A too early, as is shown in Fig. $14K$. At A the control grid becomes

* The time constant of $6C$ and $21R$ is about $\frac{1}{8}$ sec, so the voltage across $6C$ changes little during zone G.

positive while the third grid of tube 11 is still positive; with both grids positive, tube 11 passes current and electrons flow from point 8 through tube 11 and 15P to point 62. This sudden lowering of the positive or 15P side of 9C also drives the tube-10A cathode far more negative than point 8 or the 5P slider. Tube 10A passes current so that its anode 103 also is pulled negative; this negative pulse passes through a resistor-capacitor network that holds terminal 83 negative for a time long enough to let the correction motor reduce the web speed.*

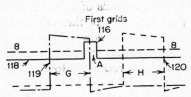

FIG. 14K. Grid potentials, spot arriving early.

* Even though the register mark is large or is moving slowly, thereby making tube 11 pass current for, say, $\frac{1}{10}$ sec, the short time constant ($\frac{1}{100}$ sec) of 9C and 18R lets cathode 114 of tube 10A recover or rise quickly after tube 10A has lowered its anode-103 potential. Therefore to provide a correction signal that is not affected by the size or speed of the register mark, tube 10A passes

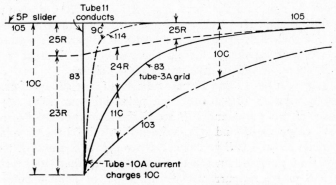

FIG. 14L. Circuit action at left in Fig. 14I.

current for perhaps only $\frac{1}{1000}$ sec, to charge capacitor 10C. Although charged so quickly, 10C discharges more slowly though 23R and 25R ($\frac{1}{3}$ sec time constant). Moreover, to provide a large correction signal for an instant at grid 83 of tube 3A, followed by a lesser signal for a longer time, capacitor 11C and 24R are added ($\frac{1}{10}$ sec T.C.). As is shown in Fig. 14L, the tube-10A current at once pulls point 103 far negative; since 11C cannot charge at once, grid 83 also is pulled far negative for the first instant. As 11C charges through 24R to the voltage across 23R, the grid-83 potential rises quickly until it remains negative only because of the voltage across 25R. When 10C has discharged, grid 83 returns to the potential at the 5P slider, ready for the next correction signal.

When the register mark arrives too late, impulse A makes the first grids of tubes 11 and 12 positive while the third grid of tube 12 is also positive. Current passes through tube 12 and $15R$; for an instant the charge on $8C$ drives negative the tube-$10B$ cathode. In the same manner as is described above, the potential at terminal 82 is driven negative, and the tube-9 current increases (in Fig. $14E$) so that the correction motor raises the draw-roll speed and advances the web.

CHAPTER 15

THY-MO-TROL*—AUTOMATIC TUBE CONTROL OF D-C MOTORS

Many types of general-purpose tube-controlled d-c motor equipment have seen years of useful service on machine tools and other industrial drives. Several sizes and kinds of Thy-mo-trol have been described in earlier texts;* more recent designs are described here.

FIG. 15A. Half-wave Thy-mo-trol panel for ½-hp motor (CR7507–F170).

15-1. A Small Half-wave Thy-mo-trol Drive (CR7507–F170). To convert an a-c supply into power suitable for a d-c motor, most equipments include two or more rectifier tubes in the armature circuit, as are shown in Fig. 13D and described in Sec. 13-3; these

---

* Thy-Mo-Trol is the General Electric Company's registered trade mark for its electronic motor control. See Chute, "Electronics in Industry," Chaps. 24 and 25; also Cockrell, "Industrial Electronic Control."

tubes rectify or use both half waves of the alternating current. To provide a variable-speed motor control with very few tubes, Fig. 15A shows a control for a ½-hp motor, wherein a single thyratron tube supplies all the power for the armature; a second tube supplies d.c. for the motor field and the control circuits. One speed dial (mounted remote) may adjust the motor speed through a range of 20 to 1, or, for example, from 1750 rpm down to 78 rpm.

The diagram of this simple control appears in Fig. 15C. At the top of this diagram, a-c power is applied to the correct tap of transformer $1T$ so as to give rated filament voltage for heating tubes 1 and 2. A start-stop push button picks up contactor M to start the motor. As is shown in Fig. 15B, power is applied to the motor armature during the shaded parts of the a-c wave; the negative half waves, shown below the 0 line, never are used.*

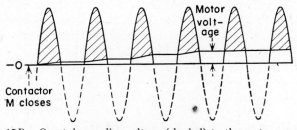

Fig. 15B. One tube applies voltage (shaded) to the motor armature.

Since all of the shaded half waves have the same polarity, they force current through the motor armature always in the same direction, as is required by any d-c motor. In Fig. 15C these electrons flow from line 1 up through the motor armature (and series or commutating fields), through starting contactor M and overload relay OL, from cathode (heated filament) to anode of tube 1 and through the fuse to line 2. These electrons flow only during the half waves when line 2 is more positive than line 1. Meanwhile the motor receives constant field current; tube 2 rectifies the a-c voltage of $S1T$ so that electrons flow from center tap 14 up through $4R$ to 1, through the field winding and tube 2 to $S1T$.

As the motor speed rises, Fig. 15B shows that the motor armature produces a d-c voltage of its own (called the counter electromotive force or back emf); the amount of this motor voltage is a true or

* The wave shapes in Fig. 15B are shown as though the motor armature has no inductance. The true wave shapes appear in Fig. 15D.

accurate measure of the motor speed. Although tube 1 may fire during the shaded half waves, current will flow into the motor armature only while the shaded wave is higher (more positive) than the motor voltage.*

When the stop button or the overload contact OL drops out relay M, disconnecting the motor armature from tube 1, an M contact (n-c) connects the motor armature to $5R$, which acts as a dynamic-braking resistor. The motor now acts as a generator (since it still has full field strength) and its back emf forces current through $5R$; this load slows the motor quickly.

The motor shunt field receives about 230 volts d.c. through tube 2; in passing also through $4R$, the field current produces a small voltage across $4R$ so that point 14 is about 5 volts more negative than point 1. The speed adjuster $1P$ may be set so that its slider-20 potential is anywhere from -5 volts to $+125$ volts, compared with point 1. When $1P$ is turned to higher potential, the motor speed increases until its back emf or motor voltage is nearly the same as the voltage from point 20 to 1.

To see how tube 1 accurately controls the motor speed, we must study its grid-to-cathode voltage. While contactor M is picked up, to run the motor, notice that the tube-1 cathode 5 may be at nearly the same voltage as point 1 if the motor is turning very slowly; at a greater speed, the back emf or motor voltage keeps cathode 5 perhaps 100 volts higher than point 1. For a moment let us connect together points 20 and 23 in the tube-1 grid circuit, so as to remove the effect of the $S1T$ voltage. If $1P$ now is set for medium speed so that slider 20 is only 80 volts above point 1, the grid of tube 1 is more positive than its cathode so long as the back emf or motor voltage is less than 80 volts. Until the motor comes up to that speed where it produces 80 volts of back emf, tube 1 fires and applies greater voltage from the a-c supply, to raise the motor speed. However, when the motor reaches higher speed so that it produces, say, 85 volts, the tube-1 grid (at 80 volts) now is 5 volts more negative than the cathode, so tube 1 does not fire; the motor coasts or slows down until it produces closer to 80 volts, so that tube 1 again may fire.

15–2. A 7-volt Wave Regulates Tube Current. Such on-and-off action of tube 1 does not give smooth or close speed control, therefore let us remove the 23-to-20 connection mentioned above, so

---

* This statement omits the effect of the armature inductance shown in Fig. 15D.

as to see the effect of $S1T$, $2C$ and $2R$ at the center of Fig. 15C. The a-c voltage between points 20 and 23 is a 7-volt wave that lags 90 deg behind the $S1T$ voltage and the anode voltage of tube 1,

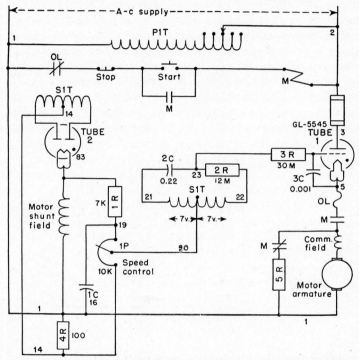

Fig. 15C. Circuit of half-wave Thy-mo-trol drive (CR7507–F170).

because the ohm value of $2R$ is about equal to the inductive-ohm value of $2C$.* This 7-volt a-c wave is shown in Fig. 15D; it swings below the $1P$-slider-20 potential during the early portion of the

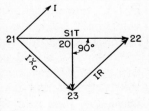

* To learn the effect of $2C$ in an alternating-current circuit, recall that the reactive ohms of a capacitor $= 1/2\pi fC$(in farads). Since $f = 60$ cycles, the reactive ohms $= 2650/C$(in mu f), so the 0.22 mu f capacitor $2C$ has $2650/0.22 = 12,045$ ohms impedance. This is nearly equal to the 12,000 ohms of $2R$. In the vector diagram of this circuit, the voltage IX_c across $2C$ has the same length as the voltage IR across $2R$. The 15-volt $S1T$ winding has a center tap 20; the voltage from 20 to the junction 23 is seen to be half the length of $S1T$, and in a direction 90 deg behind $S1T$.

positive half wave of a-c supply voltage. Meanwhile the potential of the tube-1 cathode 5 is close to 20 and is held more positive than point 1 because of the back emf of the turning motor. At point U the 7-volt wave rises close to the tube-1 cathode potential 5 and tube 1 fires; at once the wave of a-c supply voltage is applied through tube 1 and appears mostly across the motor armature, causing a flow of current that turns the motor. At point V the a-c supply voltage becomes less than the back emf of the motor; however, current continues to flow a little beyond V because of the inductance ("electrical inertia") of the armature circuit. When the motor armature current stops, tube 1 disconnects the a-c supply voltage from the motor; the back emf of the motor appears again at W.

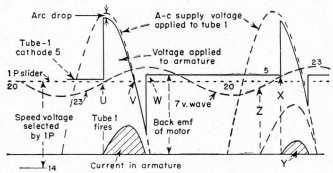

Fig. 15D. Action in circuit of Fig. 15C to regulate motor speed.

Suppose that the motor speed has risen too high so that, at W in Fig. 15D, the back emf of the motor has become 2 volts greater than before; meanwhile the 1P-slider-20 potential has not changed and the 7-volt wave appears as before. The 2-volt increase in back emf raises the cathode 5 potential by 2 volts; tube 1 is not fired until, at X, the 7-volt wave has risen 2 volts higher (than before at U). Since tube 1 is fired later in the half wave of a-c supply voltage, the pulse of armature current is smaller, at Y, so that the motor speed may decrease to the amount selected by the 1P slider.

Similarly, if a load suddenly is applied to the motor, its speed decreases. However, a slight drop in speed immediately decreases the back emf and lowers the cathode-5 potential. The 7-volt wave now rises near cathode 5 much earlier (as at Z) so that tube 1 is fired early and forces a large pulse of current through the motor armature, to bring the motor back to the desired speed.

Because of the 7-volt a-c wave added by $S1T$, $2C$ and $2R$, a small change of motor speed or load can change the armature current by a large amount, so as to hold the speed more nearly constant. When the selected speed gives a back emf of 80 volts, the motor receives no current from tube 1 when its emf is 83 volts, but the motor receives full current at 77 volts; the circuit regulates to prevent a change of more than 1 or 2 volts.

In this half-wave Thy-mo-trol drive, the motor current flows in pulses each less than $\frac{1}{120}$ sec long. Between these pulses, the motor acts as a tachometer generator to give an accurate speed signal; 60 times each second, the motor's back emf newly sets the cathode-5 potential of tube 1, which then is fired earlier or later to supply the correct current to hold the motor at the speed selected by $1P$.

15-3. A Full-wave Thy-mo-trol System. A similar all-tube control for a larger motor uses two or more thyratron tubes in the

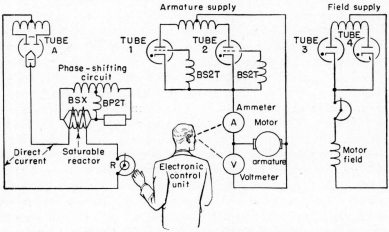

Fig. 15E. Basic arrangement of thyratron control of a motor.

armature circuit and two tubes for the motor field; this arrangement is shown in Fig. 15F, together with numerous small high-vacuum tubes in the upper or "brain" section, to be explained later.

A simple approach to this full-wave thyratron motor control is shown in Fig. 15E. The armature of the d-c motor receives its current through thyratron tubes 1 and 2, while the motor field is supplied by tubes 3 and 4. The man watching the meters is like the "electronic control unit" or "brain" of the system, described later.

232 ELECTRONIC MOTOR AND WELDER CONTROLS

Thyratrons 1 and 2 are phase-shifted by the a-c voltages of grid-transformer windings $BS2T$, so as to change the average voltage across the motor armature. The primary $BP2T$ is in a phase-shifting bridge, controlled by the saturable reactor BSX. This reactor is controlled by the amount of direct current flowing in its d-c winding; in Fig. 15E, this direct current depends on the setting of rheostat R.

Tubes 1 and 2 control the voltage applied to the motor armature in much the same way as a motor-generator set (with Ward Leonard field control) supplies variable voltage to a d-c motor. That is, if rheostat R is shorted and permits maximum direct current to flow and saturate BSX, then tubes 1 and 2 supply maximum voltage to the motor armature, and high-speed results. However, when R is turned so as to decrease the direct current in BSX, tubes 1 and 2 supply less average voltage to the motor armature, whose speed decreases.

The man in Fig. 15E, who watches a voltmeter and ammeter in the motor-armature circuit, turns rheostat R so as to hold the desired motor voltage and speed and keep the motor current within safe limits. In the "brain" of the Thy-mo-trol drive several tube circuits are used in place of this man, but the results are the same: (1) They "watch" the voltage across the motor armature; if the voltage drops even less than a volt, they cause an instant increase of direct current in BSX, making tubes 1 and 2 fire earlier in their half cycles, until the armature voltage returns to normal. (2) They "watch" the amount of motor-armature current; when that current reaches the preset limit (such as 150 per cent of rated full-load current) the direct current in BSX is decreased so that tubes 1 and 2 fire later, preventing any further increase in armature current.

For a wider range of motor speed than can be obtained from changing the armature voltage alone, the Thy-mo-trol drive may include grid-controlled thyratrons for the field supply, as well as for the armature supply. Circuits are added in the "brain" to control these field-supply thyratrons. In the complete circuit next to be studied, we shall see that these added field circuits may be quite like the armature-control circuits.

15–4. A 2- or 3-hp Thy-mo-trol Drive (CR7507–G270).* The entire elementary diagram is shown in two parts. Figure 15I

---

* The diagrams show the circuit arrangement used with a 2-hp 230-volt d-c motor. The same equipment is used to control a 3-hp 350-volt motor, by changing several resistor values.

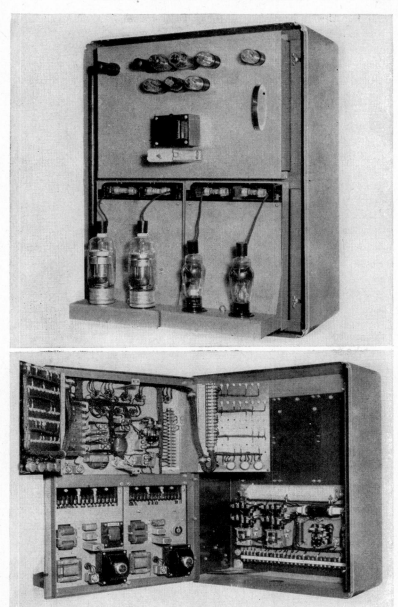

Fig. 15F. A full-wave Thy-mo-trol panel for 3-hp motor (CR7507–G270).

includes the motor with its armature- and field-supply tubes, phase-shifting bridge circuits and control station. The electronic control unit, or "brain," appears in Fig. 15J, including all tubes marked by letter. Notice the five wires that bring armature-voltage and field-current signals from the motor into the "brain"; transformer secondaries $SA1T$ and $S3T$ appear in Fig. 15J, while their primaries are in Fig. 15I.

When the a-c supply circuit is closed, transformer $1T$ applies anode voltage to thyratron tubes 1, 2, 3 and 4; a winding $S1T$ furnishes 115 volts for the control-station circuits (right-hand side of Fig. 15I). Transformers $A1T$, $B1T$ and $C1T$ furnish filament voltage to heat the tubes. The tap switch at one end of $PB1T$ must be set so that the thyratrons receive correct filament voltage, to give proper operation and long tube life.

Most parts of Fig. 15J operate on direct current, supplied by rectifier tube A. Let us first study the left-hand portion of this "brain," to see how it regulates the motor-armature voltage; these circuits are shown alone in Fig. 15G. Here we recognize tube A and the filter ($1C$ and $1X$), that provide a d-c supply between points 4 and 1. As is described in Sec. 2–4 and Fig. 2D, the voltage-regulator tubes B and C provide a more constant voltage between points 3 and 1. Because of these tubes we may think of point 2 as a d-c bus that is always 150 volts more positive than point 1; point 3 is always 300 volts above point 1, or 150 volts above point 2. Between 3 and 1 in Fig. 15G is a voltage divider ($4R$, $5R$, $6R$); notice that this divider is not connected to point 2.

15–5. Armature-voltage Control. In Fig. 15G, BSX is the d-c winding of the saturable reactor that controls the phase-shifting of armature thyratrons 1 and 2. This BSX d-c winding is in series with tube $D1$, so tube $D1$ controls the amount of direct current flowing in BSX and, therefore, acts as does rheostat R shown in Fig. 15E. From now on, remember that the armature voltage decreases when BSX current decreases. See now how tube $D1$ is controlled by tube $E1$.

When no anode current passes through tube $E1$, the voltage drop across resistor $4R$ is caused entirely by current flowing through $5R$ and $6R$. Under this condition, the resistances of $4R$, $5R$ and $6R$ have been selected so that point 10 (grid of tube $D1$) is near the same potential as point 2 (cathode of tube $D1$) so that tube $D1$ passes maximum current through BSX (which lets tubes 1 and 2 apply maximum voltage to the motor armature). If we now permit tube $E1$ to pass current (whose electrons flow from point 2 through tube $E1$ to point 9, through $4R$ to point 3), this current increases the voltage drop across $4R$. When this happens, there must be a corresponding decrease in drop across $6R$; this causes the potential at grid 10 to drop, so tube $D1$ decreases its current flow through BSX. Briefly, turning on tube $E1$ turns off tube $D1$ and thereby decreases the armature voltage.

Tube $E1$ is the comparison tube, which is told what armature voltage (speed) is desired, by the position of the $1VR$ slider, connected through $13R$ to its grid; it then "watches" the motor armature voltage to see that it is correct. For lowest motor speed, the $1VR$ slider touches at the top of $1VR$ in Fig. 15G, so grid 14 of tube $E1$ is near cathode-2 potential; tube $E1$ passes current, but tube $D1$ passes so little current that the motor armature receives very little voltage from thyratrons 1 and 2, and the motor barely turns.*

To increase the motor speed, the $1VR$ slider is turned downward in Fig. 15G.† Suppose that this lowers the grid-14 potential by 20 volts; this decreases the tube-$E1$ current, increases the tube-$D1$ current, turning on thyratrons 1 and 2 to apply greater average voltage to the armature so that the motor speed rises. Most of this motor voltage is filtered or smoothed by capacitor $4C$, and is applied across the voltage divider ($A2VR$, $12R$, $13R$); each 4-volt rise at armature terminal 45 causes about 1 volt rise at grid 14. When the motor-armature voltage has increased to 80 volts, nearly 20 volts appears across $13R$ so that the grid-14 potential is returned close to cathode-2 potential; tube $E1$ again passes enough current so that tube $D1$ is partially turned off, and thyratrons 1 and 2 apply just enough voltage to keep the motor turning at the speed produced by 80 volts. If $1VR$ again is set for higher or base speed so that the $1VR$ slider is 75 volts more negative than cathode 2, tube $E1$ again is turned off and thyratrons 1 and 2 are turned on completely, to supply as much as 300 volts

---

* Above $1VR$ in Fig. 15J, another dial adjusts this minimum speed; its resistance prevents the $1VR$-slider potential from rising too close to cathode 2.

† In Figs. 15G and 15H the speed potentiometer $1VR$ is shown connected to bottom point 1. However, as appears in Fig. 15J, a resistor $1R$ is inserted between $1VR$ and point 1. Therefore, when $1VR$ is turned downward for higher motor speed, the $1VR$ slider can become only 75 volts more negative than point 2.

to the armature.* Again the voltage across 13R (about ¼ of 300 volts) raises grid 14 close to cathode 2; thyratrons 1 and 2 continue to supply full armature voltage to the motor so that it runs at base speed.

In the grid circuit of tube E1 in Fig. 15G, we say that a portion of the motor-armature voltage (across 13R) is compared with or is balanced against a reference voltage selected by 1VR (from 2

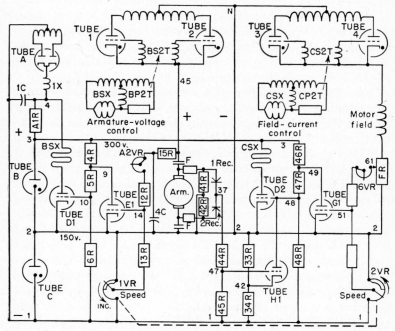

Fig. 15G. Armature and field-control portion of circuit shown in Figs. 15I and 15J.

to 1VR slider). If a dip in supply voltage causes 1/10 volt drop in armature voltage, it also causes a tiny drop at tube-E1 grid; meanwhile cathode 2 and the 1VR slider remain at constant potentials (thanks to the voltage-regulator tube C). This tiny voltage unbalance causes instant action by the tubes, to return the armature voltage to normal.

* The resistance in $A2VR$ may be decreased to lower the maximum value of this voltage; a larger portion of the armature voltage then is applied across 13R.

15–6. Field-current Control. Similar to the armature-voltage-control circuit described above, the motor field-current-control circuit appears at the right in Fig. 15G, and operates from the same d-c supply (points 1, 2 and 3). It has the voltage-divider 46R, 47R, 48R. Tube $G1$ balances a field-current signal (between FR tap 61 and point 2) against the desired field current as set by $2VR$. At base speed or below, the $2VR$ slider is at the negative potential of point 1; grid 51 of tube $G1$ is quite negative so tube $G1$ draws little or no current through 46R. Tube $D2$ passes enough current through saturable reactor CSX so that thyratrons 3 and 4 furnish full current to the motor field.* The potential at FR tap 61 becomes more positive than point 2, so that grid 51 of tube $G1$ is perhaps 5 volts more negative than its cathode 2.

When the $2VR$ slider is turned upward (above its middle position corresponding to base speed) so as to raise the potential at grid 51, the tube-$G1$ current increases and causes a larger voltage drop across 46R; the lowered potential at 48 decreases the current of tube $D2$ through CSX, and thereby decreases the current that tubes 3 and 4 supply to the motor field. The potentials at 61 and grid 51 are thereby lowered to reach a balance.

Notice that $1VR$ and $2VR$ are mounted on one shaft and are operated by a single dial. In Fig. 15G the lower half of $1VR$ is a dummy section of almost zero resistance; the same applies to the lower half of $2VR$. Turned clockwise, these potentiometers cause lowest motor speed, for the armature voltage is almost zero but the field current is large (tube $E1$ is full "on," tube $G1$ is all "off"). Turning these potentiometers counterclockwise gradually increases the armature voltage but has no effect on the field current until the mid-point is reached; at mid-point the motor operates at base speed — full armature voltage and full field. (Here the motor is assumed to have 2 to 1 speed range by field control; that is, its top speed is twice its base speed.) Further turning has no effect on armature voltage but gradually decreases the field current, thereby causing further increase in motor speed.†

* The maximum field current is adjusted by selecting the tap on resistor FR and by setting 6VR. The resistance of FR does not change with temperature, so the voltage across FR holds constant the motor-field current despite changes in field resistance.

† Notice that 7VR in Fig. 15J adjusts and limits how far the grid potential of tube $G1$ can be raised, and thereby limits the maximum motor speed. Also, resistor 39R is placed around tube $D2$ to make sure of enough field current to prevent too high speed if tube $D2$ should fail.

15–7. Preventing Overvoltage. Near the center of Fig. 15G is an overvoltage control that acts mainly when the speed dial is turned quickly from a high-speed to a medium-speed position, strengthening the motor field. Here the motor acts as a generator and tries to "pump back" into the line. However, the rectifier (one-way) action of tubes 1 and 2 prevents such pump-back, therefore the voltage generated by the motor armature can become too high. To prevent this overvoltage, tube $H1$ is added. Cathode 42 of tube $H1$ is held about 90 volts above point 1; its grid 47 connects through $44R$ and rectifier 1Rec so as to respond to the armature voltage.*

As long as the armature voltage is less than 300 volts (for a 230-volt motor), the potential at grid 47 is so far below cathode 42 that tube $H1$ passes no current. However, when excess armature voltage occurs, the increased voltage across $41R$ raises point 37 and grid 47 so that electrons flow from cathode 42 through tube $H1$ to 48, through $47R$ and $46R$ to 3. This lowers the grid-48 potential so that tube $D2$ decreases the current in CSX; this reduces the field current (although it opposes the action of $2VR$ and tube $G1$) so that the voltage generated by the motor is held within safe limits.

15–8. Current Control. All the armature-control circuits this far have acted like the operator in Fig. 15E, while he looks only at the voltmeter. To let this "brain" control armature current, something must act like the ammeter in Fig. 15E, to indicate the amount of armature current. The armature-current-control section of Fig. 15J is shown separately in Fig. 15H. In these diagrams a primary winding $P3T$ is placed above tube 1, and a duplicate $P3T$ above tube 2. These are windings of a current transformer, whose secondary $S3T$ appears above tube F. When current flows (one half cycle through tube 1, the next half cycle through tube 2), a corresponding amount of alternating current flows in $S3T$; tube F rectifies this current into a pulsating direct current, so that electrons flow from the $S3T$ center tap to 33 and

* At the center of Fig. 15G, resistors $41R$ and $42R$ and rectifiers 1 and 2 are part of the group shown at the lower left in Fig. 15I. When the F contacts are closed, about $3/5$ of the armature voltage appears across $41R$ (with no voltage across $42R$); 1Rec connects point 37 to $40R$ and the positive armature terminal. When the R contacts close and the motor turns in the reverse direction, voltage appears across $42R$ (but not across $41R$) so 2 Rec connects point 37 to $43R$ and the positive armature terminal. No matter which way the motor runs, the armature voltage makes point 37 more positive than 2.

up through $5VR$, $25R$ and tube F. When no current flows through tubes 1 and 2 or the motor armature, no current is flowing in $S3T$, tube F, $25R$ or $5VR$; points 33 and 26 are at the potential of point 2.

15-9. Limiting the Motor Current. Before armature current flows, the tube-$E2$ grid is more negative than cathode 2, for its voltage divider $23R$–$24R$ is connected between 26 (now at 2 potential) and the negative $4VR$ slider. When current flows through

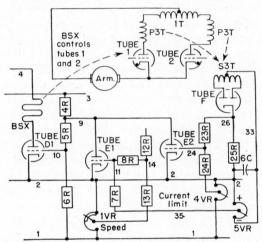

Fig. 15H. Armature-current-control portion of circuit shown in Figs. 15I and 15J.

$S3T$ and tube F (increasing as the armature current increases), this current produces a voltage drop across $25R$, raising the potential of point 26. If $4VR$ is set to permit, say, 130 per cent of rated armature current, the rise of point-26 potential also raises the grid-24 potential, but not high enough to turn on tube $E2$. However, more than 130 per cent current raises grid 24 enough to let current flow through tube $E2$ and $4R$; this lowers the potential at tube-$D1$ grid 10 and decreases the current through BSX so that the armature current also decreases or is limited to 130 per cent.

If $4VR$ is turned clockwise to a lower potential, perhaps 150 per cent current may flow before points 26 and 24 are raised high enough to limit the current.

15-10. Constant Speed with Changing Load. The thyratron-control system has the ability to keep the motor at constant speed, regardless of changes of motor load. This is more than is expected of a standard d-c motor operating from a constant-voltage supply. Even with constant armature voltage, we know that the motor speed usually changes when the motor load changes, because of the internal voltage drop (IR drop) of the motor. However, if we can increase the armature voltage a certain amount to match each increase in load, the motor speed can be held more constant. The Thy-mo-trol drive includes a speed-drop adjuster (also called the "IR-drop compensator"); by proper adjustment of potentiometer $5VR$ in Fig. 15H the armature voltage is raised just enough to offset the motor's natural desire to slow down under the increased load, as is explained next.

In Sec. 15–5 we showed that when grid 14 of tube $E1$ is made more negative, thyratrons 1 and 2 increase the motor speed. In Fig. 15H resistors $7R$ and $8R$ are added as a voltage divider between point 14 and the slider of $5VR$; point 11 on this divider now is connected to the grid of tube $E1$. When the $5VR$ slider touches at point 2, the tube-$E1$ grid is slightly more negative than 14 and responds only to the changes of armature voltage; it maintains constant armature voltage, regardless of motor load. However, if $5VR$ is turned clockwise, the voltage between 2 and the $5VR$ slider has an added effect on tube $E1$.

The voltage across $5VR$ increases as the motor load increases. Notice that the current through $S3T$, tube F and $5VR$ makes the slider of $5VR$ more negative than the top, point 2. This $5VR$ voltage contains ripples (for it is merely rectified a.c.), so the large capacitor $6C$ acts as a filter to remove the ripple from the voltage across $5VR$.

As the motor load increases and the voltage across $6C$ increases, the potential at $5VR$-slider 35 and grid 11 of tube $E1$ is lowered also; tube $E1$, acting through tubes $D1$, 1 and 2, causes the armature voltage to increase. By proper setting of speed-drop adjuster $5VR$, this increased armature voltage will hold constant motor speed, despite changes in load.

15–11. Starting the Motor. When the a-c power switch is closed, relay TR (at the right in Fig. 15I) times the 1-minute period needed for warming the tubes. During this time one TR contact disconnects one side of saturable reactor CSX; this prevents grid transformer $C2T$ from firing tubes 3 and 4 early enough to pass any field current. When this TR contact closes, field current flows through tubes 3 and 4 (and picks up field-loss relay

FLR, if used); another TR contact closes in the push-button circuit. If overload relay OL is closed, the motor may be started by either the "Forward" or the "Reverse" button. To reverse a direct-current motor, the Thy-mo-trol drive uses "forward" and "reverse" contactors in the armature circuit, shown as F and R.

When the "Forward" button picks up contactor F, the button also closes a circuit to points 14 and 15 and through F contacts to 19 and through 5Rec and $G1R$ to pick up $1CR$. Relay $1CR$ closes its contact 17–19 to seal in contactor F; two F contacts connect the motor armature (at the lower left in Fig. 15I) to positive terminal 45 and negative terminal 2 of the rectifier (tubes 1 and 2). The motor speed rises to the amount set by the speed dials ($1VR$ and $2VR$ in Fig. 15J); even if these dials are set at a high-speed position, the motor comes to this speed without drawing more current than is preset by the "Current-limit" dial $4VR$. Similarly, if the motor is started by the "Reverse" button, contactor R is picked up; this button also closes a circuit to points 15 and 14 and through R contacts, 5Rec and $G1R$ to pick up $1CR$, whose contact 17–19 seals in R. When either F or R is picked up, relay $2CR$ also is picked up, to open its contacts as mentioned below.

With the arrangement shown in Fig. 15H, the motor may be sluggish or slow to start if $1VR$ is set in a low-speed position; both the $1VR$ and $5VR$ sliders are so near point-2 potential that tube $E1$ is turned on, reducing the tube-$D1$ current and the armature voltage to less than the amount needed to start the motor. To make sure of enough starting torque at any speed setting, $26R$ and a $2CR$ contact are added below $5VR$ at the bottom of Fig. 15J. Before starting the motor, this $2CR$ contact lowers the potential at all parts of $5VR$; capacitor $6C$ charges to this voltage across $5VR$. Therefore, for perhaps $\frac{1}{20}$ sec after the "Forward" or "Reverse" button is closed to start the motor, the $5VR$-slider potential holds negative the tube-$E1$ grid so that tube $E1$ does not limit the starting performance of the motor.*

Although tube $E1$ is thus turned off at the instant of starting the motor, we shall see that tube $E2$ is turned on enough to limit

---

* Before the motor is started, and while the tube-$E1$ grid is held negative at slider 35 of $5VR$, the antihunt capacitor $2C$ above tube $E1$ is charged by part of the 9-to-35 voltage; this $2C$ voltage permits an undesired overshoot of armature voltage during starting. Therefore, capacitor $15C$ is added between tube $E1$ and $5VR$ to buck the $2C$ voltage and provide smoother starting.

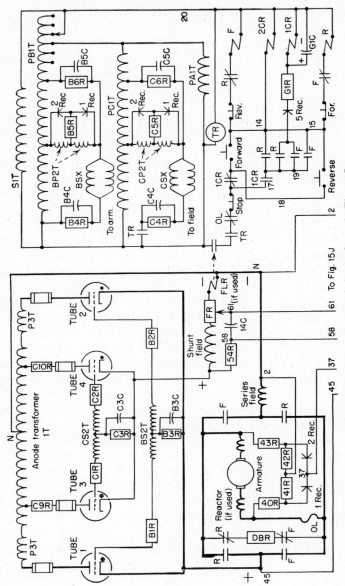

Fig. 15I. Upper portion of circuit shown in Fig. 15J.

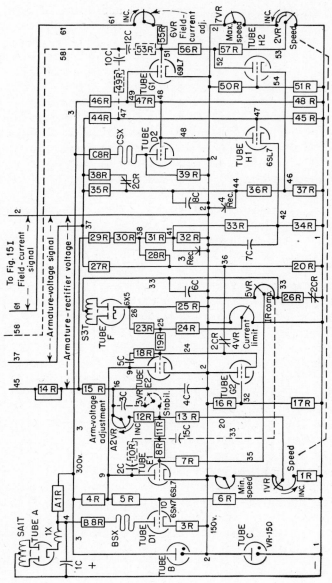

Fig. 15J. Circuit of 3-hp Thy-mo-trol drive (CR7507–G270).

the voltage available from thyratrons 1 and 2; in this way, the armature rectifier is "preconditioned" so that it supplies only, say, 50 volts at the instant when the F or R contacts connect this rectifier to the motor armature. This action limits the inrush or starting current to a safe value. Near the center of Fig. 15J, notice tube $G2$ and the $2CR$ contact near it. Before either button is pushed, this $2CR$ contact "preconditions" the circuit by connecting the tube-$E2$ grid 24 to point 36 and through $28R$ to point 38 (on voltage divider $29R$–$30R$–$31R$–$32R$ across the armature rectifier voltage).* Even if this armature-supply voltage (45 to 2) is only 50 volts, point 36 may be more positive than cathode 2 of tube $E2$. The circuit through this $2CR$ contact keeps tube $E2$ passing enough current so that tube $D1$ is nearly turned off; thyratrons 1 and 2 are phased nearly off, supplying only 50 volts at point 45.

When starting the motor, contact $2CR$ disconnects the tube-$E2$ grid 24 from point 36. The potential of grid 24 now may drop, for it is connected through $19R$ and $24R$ to the slider of $4VR$. Until motor-armature current flows, $S3T$ and tube F produce no voltage across $23R$ and $24R$, so point 25 may be as much as 150 volts below cathode 2 of tube $E2$ (depending on the setting of $4VR$). If grid 24 should approach this low potential, tube $E2$ would be turned off, tube $D1$ would be turned on, and tubes 1 and 2 would permit a large inrush of current to the motor armature. To prevent grid-24 potential from dropping too fast, we add capacitor $5C$ and $18R$ between grid 24 and anode 9 of tube $E2$.† The voltage across $5C$ must change before grid 24 can change potential; $5C$ lets the grid-24 potential drop slowly, so that tube $D1$ turns on more slowly and the armature current increases smoothly. Within $\frac{1}{10}$ sec this armature current (acting through transformer $3T$ and tube F) raises the potential at points 26 and 25 high enough to prevent further drop of grid 24; after that, the "current-limit" control at $4VR$ has full effect. While the motor is starting, its field is at full strength (see Sec. 15–12).

When the motor is running (and $2CR$ contacts are open), notice

* Resistors $27R$ and $28R$ have much smaller resistance than $20R$, so point 36 is close to point-37 or point-38 potential.

† Capacitor $5C$ and $18R$ serve also to filter or smooth the grid voltage of tube $E2$, which otherwise would include jagged wave shapes of voltage caused by the delayed firing of thyratrons 1 and 2, and applied at point 45 (to point 36 and through $2CR$ contact) or through tube F and $23R$.

how $5C$ may cause an unwanted delay if the motor load increases suddenly; this delay is offset by adding tube $G2$. Without $G2$, suppose that there is very little load on the motor; with little armature current, there is little voltage across $24R$. Grid 24 of tube $E2$ has reached the low potential of the $4VR$ slider, so $5C$ has charged to the large voltage between point 9 and the $4VR$ slider. If the motor is suddenly loaded (or if the speed-control dial is quickly turned to ask for higher speed), the armature current can rise too high before tube $E2$ can be turned on to limit the armature current; this is because grid 24 has been at too-low potential and can rise to turn on tube $E2$ only after $5C$ has lost much of its charge. However, tube $G2$ prevents grid 24 from reaching such low potential. Since grid 32 of $G2$ is held about 10 volts below point 2, $G2$ has no effect as long as grid 24 remains near or above point 2. But when grid 24 drops to about 10 volts below 2, the cathode of $G2$ is near the potential of grid 32 so that tube $G2$ passes current; these electrons flow from bottom point 1 through $4VR$ slider, $24R$ and $19R$ to point 24, through tube $G2$ to point 2. Enough voltage drop appears across $19R$ and $24R$ so that grid 24 is held within 10 volts below cathode 2—just low enough to cut off the current flow of tube $E2$. When stopping or reversing the motor, the $2CR$ contact again connects grid 24 to point 36; tube $E2$ (helped by tube $G2$) turns on quickly, to prevent a large "plugging" or reversing current, as is described later.

15–12. Starting with Full Field. If the motor is started while $1VR$ and $2VR$ (bottom part of Fig. 15J) are set for high speed, the slider of $2VR$ is at high potential to let tube $G1$ pass current and thereby to decrease the field current (see Sec. 15–6). However, such weakened field reduces the motor torque available for bringing the load up to speed, therefore tube $H2$ is added (at the right in Fig. 15J) so as to maintain full field until the motor reaches medium speed.

The cathode of tube $H2$ is connected to point 46 on a voltage divider $35R$–$36R$–$37R$ that receives voltage at 37; point 37 rises as the armature voltage increases. At low speeds the voltage 37-to-2 is so small that point 44 on the divider tries to fall to a potential below that of point 2. However, rectifier 4 passes current when 44 is more negative than 2, and holds point 44 close to 2. So, until the motor reaches medium speed, cathode 46 of tube $H2$ remains about 110 volts above point 1. Meanwhile, grid 54 is held at 114 volts; with this $+4$ grid voltage, tube $H2$ conducts, so that electrons flow from 46 through tube $H2$ and $57R$ to the positive potential of the $2VR$ slider (but only when $2VR$ is set for high speed). In this way point 52 is held 40 volts below point 2; grid 51 also is held at low potential, nearly turning off tube $G1$, and thereby causes nearly full current through tube $D2$, CSX and in the motor field.

When the motor approaches base speed (so that its armature voltage exceeds 90 per cent of its full amount), point 37 is raised high enough to pull point 44 more positive than 2; rectifier 4 no longer passes current. As 44 rises above 2, cathode 46 rises also, decreasing the tube-$H2$ current through $57R$. The final 10 per cent increase in armature voltage raises points 37, 44 and 46 enough to turn off tube $H2$, letting grid 51 rise and turn on tube $G1$; the field is weakened to produce the high motor speed selected by $2VR$.

15–13. Slow-down by Dynamic Braking. Pushing the stop button in Fig. 15I drops out both F and R contactors; this disconnects the motor armature from point 45 and tubes 1 and 2, so the motor coasts to a stop. However, other F and R contacts now connect the armature across a braking resistor; the fast-turning motor acts as a d-c generator, forcing current through DBR. This quickly brings the motor to rest. The amount of this braking, or slow-down, depends on the amount of DBR resistance and the strength of the motor field; with no field current, the turning motor cannot generate—it merely coasts.

While the motor is being slowed by dynamic braking as just described, the motor will stop most quickly if the motor-field current is increased so that the turning armature generates full voltage. However, if the field is strengthened too much before the motor slows to base speed, too high voltage may be produced. In Sec. 15–7 we showed how tube $H1$ acts to weaken the motor field and prevent overvoltage when the speed dial is turned quickly to a lower speed position; tube $H1$ is turned on because point 37 is raised high by the slowing motor. However, in Sec. 15–7 the F or R contacts remain closed, so that the armature voltage appears across $40R$ and $41R$, or only across $42R$ and $43R$; about 60 per cent of the armature voltage appears between points 37 and 2.

When the stop button drops out both contactors F and R, the armature is disconnected from point 2; now the armature voltage is applied across $40R$, $41R$, $42R$ and $43R$ in series, and only 30 per cent of the armature voltage appears between points 37 and 2. This lesser voltage cannot raise grid 47 high enough to let tube $H1$ protect against overvoltage. Moreover, this decreased potential at 37 lets all parts of the divider $35R$–$36R$–$37R$ fall quickly to lower potential, lowering cathode 46. If $2VR$ is set in a high-speed position, the tube-$H2$ current helps to turn off tube $G1$, thereby strengthening the motor field too quickly.

To prevent this undesired action, a $2CR$ contact connects $38R$ across $35R$ (near tube $D2$ in Fig. 15J), whenever both contactors F and R are dropped out. Although the voltage 37-to-2 is only half of its previous value, the decreased resistance between 37 and 44 now causes a larger portion of the 37-2 voltage to be applied across $36R$ and $37R$; cathode 46 is held high enough to hold off tube $H2$. So, while the motor turns at high speed, the motor field current is just enough to permit the motor to produce rated armature voltage. Moreover, as the motor speed decreases toward base speed and the 37–2 voltage becomes less, the potential at cathode 46 is lowered, gradually turning on tube $H2$ so that the motor-field current increases and the turning armature continues to produce rated voltage (and maximum braking torque) as it slows down. Below base speed, no further increase in field strength is permitted.

15–14. Wave Shapes of Armature Voltage. Before watching the circuit action during a sudden reversal of the motor, let us see what voltage is applied across the motor armature, at various speeds and loads. In (a) of Fig. 15K, the forward contactor F closes; at once tubes 1 and 2 fire (in Fig. 15G or 15I), very late in the half cycles of anode voltage. The heavy jagged line in Fig. 15K shows the wave shape that might appear on an oscilloscope* connected between terminals 45 and 2 of Fig. 15I—the voltage applied to the motor armature.

As the armature turns, it generates a d-c voltage (or back emf, mentioned in Sec. 15–1), which appears as a horizontal line (at L in Fig. 15K) during the time when tubes 1 and 2 are not passing current. This L voltage increases as the motor speed rises. Parts of the a-c voltage wave (from anode transformer $1T$) appear on the oscilloscope only during those parts of the cycle when tube 1 or tube 2 is firing; at such times the tubes connect the transformer to the armature—at other times the oscilloscope shows only the

---

* The oscilloscope must be arranged to indicate d-c voltage. See Chute, "Electronics in Industry," Chap. 27.

248   ELECTRONIC MOTOR AND WELDER CONTROLS

d-c voltage generated by the motor itself. In (b) of Fig. 15K the motor is turning at medium speed, so its generated d-c voltage (or back emf) may be half the height of the available 1T a-c voltage wave. At this speed, if the motor is loaded, tubes 1 and 2 must fire fairly early in each half cycle to let enough armature current flow to run the motor. At B the tube fires; however, only the voltage M can force current through the motor, for the rest of the a-c wave is opposed by the motor's generated voltage. Notice that armature current may flow, even after the a-c voltage has reversed, for the energy (stored in the armature inductance by the flow of current) keeps the tube firing until the current has decreased to zero. With less load, the tubes fire later (at C) and the armature current flows for a smaller part of each half cycle.

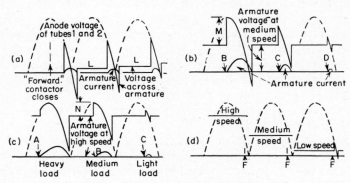

FIG. 15K. Armature-voltage wave shapes at various speeds and loads.

Without load, the motor needs so little current that the tubes need not fire until D.

At high speed, (c) of Fig. 15K shows the high position of the straight line of d-c armature voltage. Only the voltage N can force armature current to flow. If the motor carries heavy load, the tubes must fire early (at A), so that armature current flows most of the time. At lighter loads, the tubes fire later (at B or C).

If we turn the speed dial (1VR in Fig. 15J) to a lower speed position, the firing of tubes 1 and 2 is delayed until late in the half cycle, shown at F in (d) of Fig. 15K. While the motor coasts* to a lower speed, the oscilloscope shows only the horizontal line of d-c voltage; this line gradually falls to a lower level as the speed

* Faster slow-down is gained if another tube is added to the "brain" of the Thy-mo-trol drive, as is described in Sec. 15–20.

THY-MO-TROL D-C MOTOR DRIVE

decreases. When this d-c voltage reaches a low level, the tubes again fire, as shown at the far right in (d), to keep the motor turning at low speed.

15–15. Reversing the Motor; Inverter Action. With the motor running in the forward direction, suppose that the "Reverse" button in Fig. 15I is pushed; this drops out F and picks up R, while the armature still is generating voltage. In Fig. 15L, the first half cycle shows the motor running forward at part speed, and E is the voltage generated by the motor. The voltage difference N is forcing armature current to flow when tube 1 fires (shown at G).

When the "Reverse" button is pushed, contactor R connects the armature to the rectifier in the opposite direction; the motor's generated voltage now adds to the a-c anode voltage, so that the

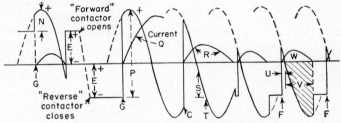

Fig. 15L. Inverter action when motor is reversed.

total voltage P will force armature current to flow when tube 1 fires. If tube 1 now is permitted to fire at this same point G, notice the large current Q, which may damage the motor. By delaying the firing point of tube 1 until C, this current is decreased. Yet this current R is much greater than the current shown in (b) of Fig. 15K, where tube 1 is fired also at this point C. With the motor reversed in Fig. 15L, current flows through tube 1 and the armature in the same direction as before (for current R is above the zero line). However, even after the a-c voltage of anode transformer 1T has become negative, voltage S still can force armature current to flow. Later than at point T, armature current flows until the energy in the armature inductance is used.

Before the "Reverse" contactor picks up, suppose that we further delay the tube-1 firing point until F, near the end of the positive half cycle of anode voltage in Fig. 15L. During V, current flows through tube 1 and the armature in the same direction as during U. However, during V, this current is flowing during the negative half cycle of anode voltage; here energy is flowing from the motor armature, through tube 1 into anode transformer $1T$ and the a-c supply line. A half cycle later, tube 2 likewise fires late at F and lets the armature-generated voltage force energy into the a-c line. In this way, this rectifier acts as an inverter, changing direct current (generated by the armature) into alternating current. This action loads the armature so that it quickly stops. At once it may turn in the reverse direction and reach the speed set by $1VR$.

15–16. Time Delay before Inverter Action. Figure $15L$ shows the need for delaying the firing point of thyratrons 1 and 2 during a reversing operation, as described above. This phase shift (or delayed firing) results when the $2CR$ contact (near tube $G2$ in the center of Fig. $15J$) connects grid 24 to point 36. Since 36 is connected through $28R$ to point 38, the firing of tubes 1 and 2 is "preconditioned" as is described in Sec. 15–11. Even this amount of phase shift may not delay the firing of tubes 1 and 2 enough to prevent high armature current during inverter action. Therefore point 36 is connected also through $27R$ to point 37; while the motor is slowing down (in its original direction), its generated d-c voltage holds point 37 at high potential so that point 36 and grid 24 are held high enough to turn off tube $D1$ completely, delaying the firing of tubes 1 and 2 until later than F in Fig. $15L$.

To produce the above results, the $2CR$ contact must stay closed long enough to turn on tube $E2$. To provide the needed delay between the opening of F contacts and the closing of R contacts (and to let the $2CR$ contacts stay closed longer), relay $1CR$ is added at the lower right in Fig. $15I$. While motor contactor F or R is picked up, relay $1CR$ is picked up also; the n-o $1CR$ contact seals in F or R, while the n-c $1CR$ contact opens in the push-button circuit.

Suppose that F is picked up and the motor is running "Forward." Relay $1CR$ also is picked up (see Sec. 15–11); owing to rectifier $5Rec$, only half waves of voltage (when 19 is more positive than 20) are applied to the $1CR$ coil. Capacitor $G1C$ is charged by these half waves. If the reverse button now is pushed, it drops out contactor F but it cannot pick up contactor R so long as $1CR$ still is picked up. Since both the F and R contacts are open between point 19 and $5Rec$, the $1CR$ coil receives no further voltage from the a-c supply; however, $1CR$ is held in for about $\frac{1}{10}$ sec while capacitor $G1C$ discharges through the $1CR$ coil. When

$1CR$ drops out after this delay, the n-c $1CR$ contact permits the "Reverse" button to pick up R; $1CR$ again is picked up, to seal in R. So, when the motor is being reversed, relay $1CR$ delays the pick up of the closing contactor, and makes sure that the 24-to-36 circuit has had time to phase back tubes 1 and 2 (to fire later than point F in Fig. 15L).

15–17. Limiting Inverter Voltage. While the motor still is turning in a forward direction, the closing of the reverse contactor now connects the negative side of the armature to terminal 45 (at the lower left in Fig. 15I); point 45 is driven more negative than point 2. To control and limit the armature voltage generated here, tube $H1$ comes into action. In Fig. 15J, the tube-$H1$ cathode 42 is connected to voltage divider $33R$–$43R$; owing to rectifier 3Rec, terminal 41 of $33R$ is held at point-2 potential while the motor runs normally (for points 45 and 38 are more positive than point 2). During reversal, however, while the motor forces point 45 more negative than 2, point 41 also is driven below 2 (since 3Rec no longer passes current to connect 41 to 2). It follows that cathode 42 is driven more negative because of this reversed armature voltage; meanwhile grid 47 of tube $H1$ receives its armature-voltage signal from point 37 as before (see Sec. 15–7, footnote). Therefore tube $H1$ passes current while the motor is slowing down; as is described in Sec. 15–7, this tube-$H1$ current decreases the motor-field current so that the voltage generated by the motor is held within safe limits.

15–18. Thyratron Grid-voltage Wave Shape for Inverter Operation. To provide successful inverter operation for reversing the d-c motor, thyratron tubes 1 and 2 must receive a grid-voltage signal that is not merely a sine wave; this is shown in Fig. 15M for a single tube. Part (a) of Fig. 15M shows a sine-wave grid voltage that is satisfactory in many circuits; although this wave fires the tube late, at Q, it does not cause unwanted firing, at P. However, this same sine wave may lose control of its thyratron in a motor-reversing circuit. As was indicated previously, at P in Fig. 15L, the closing of the "Reverse" contactor connects the turning armature so that its generated voltage is added to the half wave (of a-c supply voltage) applied across the thyratron. Part (b) of Fig. 15M shows how this motor back emf holds the tube cathode more negative than the anode, during more than half of each cycle. If the sine-wave grid voltage tries to fire the tube late, at S, it also may cause unwanted firing, at R. Therefore, instead of a sine wave, a special voltage wave is supplied to the thyratron grids in Fig. 15*I*; the circuits that produce this special wave are shown again in Fig. 15*M*. The special grid-voltage wave appears in part (*c*); although it fires the tube very late, at *U*, it does not cause unwanted firing at *T*, where the tube anode first becomes more positive than the cathode.

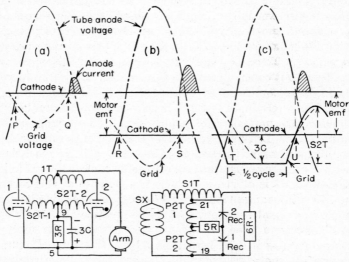

Fig. 15*M*. Forming the grid-voltage wave shape needed for inverter action.

This new grid-voltage wave is merely half of an a-c sine wave, but is located so that its horizontal line or axis is held far below the tube-cathode potential, because of the charge on 3*C*. The grid transformer has two primary windings *P2T*, wound on separate legs of the transformer iron core; the secondaries *S2T* are separated also. Although a full a-c wave is applied between points 19 and 21, only *P2T*–1 receives this voltage during the half wave when 21 is more positive than 19, for current flows through *P2T*–1, 5*R* and 1Rec; *S2T*–1 produces the grid-voltage wave that fires tube 1. During the following half cycle, when 19 is more positive than 21, *P2T*–1 receives no voltage; current flows through *P2T*–2, 5*R* and 2Rec so that *S2T*–2 produces the grid-voltage wave that fires tube 2. As each tube grid in turn becomes more positive than its cathode, the *S2T* voltage forces electrons through 3*R*, cathode to tube grid and back to *S2T*; so, by grid rectification, capacitor 3*C* is charged so as to hold the *S2T* windings at lower potential than the tube cathodes, as is shown in (*c*) of Fig. 15*M*.

15–19. Antihunt Circuits. In Fig. 15J, the tube-$E1$ grid potential is received mainly from a voltage divider (15R, $A2VR$, 12R, 13R) connected between the speed-selector voltage (1VR slider) and the motor-armature-rectifier voltage at point 45. For a stability or antihunt adjustment, capacitor 3C and 3VR are added so that 3C becomes charged to the steady d-c voltage across $A2VR$ and 12R, which is perhaps three-quarters of the 16-to-20 voltage. Slow changes of armature voltage at 45 may cause only one-quarter as much change at the tube-$E1$ grid (even if tube $E1$ fails to regulate the armature voltage). However, any sudden voltage rise at 45 causes a similar rise at both terminals of 3C; this raises the tube-$E1$ grid a large amount, forcing tube $E1$ to respond quickly to turn off tube $D1$ and decrease the armature voltage. After a part of a second (as selected by 3VR), 3C will have charged to the new voltage across $A2VR$ and 12R and has no further effect on the tube-$E1$ grid potential. Because of 3C, tube $E1$ is made to give extra response to a sudden change in armature voltage, followed by lesser response until its sudden action can take effect on the armature voltage.

In contrast, 2C and 10R connect between anode and grid of tube $E1$, to oppose or slow down the effect of any sudden changes of grid potential (see Sec. 18–1).

Across the motor field (center of Fig. 15I), 14C and 54R are connected so that 14C is charged to the entire steady voltage across the shunt field. Carried by wires 58 and 61 to Fig. 15J, most of this field voltage appears also across 10C and 12C above tube $G1$. So long as the field voltage remains steady, or changes slowly, these 14C, 12C, 10C circuits have no effect; the smaller voltage 61-to-2 is applied to divider 6VR, 55R, 56R, 57R and part of this signal causes tube $G1$ to regulate the field voltage. However, any sudden increase of field voltage raises point 61; for a short time, the charge on 14C and 12C causes a similar rise at grid 51, turning on tube $G1$ to turn off tube $D2$ and thyratrons 3 and 4 and decrease the applied field voltage. For much less time, the charge on 10C raises grid 47 of tube $H1$; the tube-$H1$ current directly lowers the grid of tube $D2$, causing similar decrease in motor-field current.

15–20. Quick Slow-down. When a d-c motor receives its power from a d-c generator or bus, the motor can "pump back" or return power to the d-c supply while it is being slowed to a lower speed setting; during such slow-down the motor acts as a generator, and its load quickly reduces its speed. However, when the motor receives its d-c supply through electron tubes, such pump back is not used to slow the motor. If the speed dial ($1VR$ and $2VR$ in Fig. 15G or 15J) is turned to a lower speed position while the motor is running, current through the armature thyratrons 1 and 2 stops;* the motor merely coasts until its external load lowers its speed to the new dial setting; tubes 1 and 2 then fire to hold the motor at this new required speed.

To provide faster slow-down when needed, tube W and a quick-slow-down relay QSR are added to the Thy-mo-trol equipment as is shown in Fig. 15N. If the speed dial now is turned to ask for lower motor speed, tube W picks up relay QSR, which picks up contactor DB; the DB contacts connect the turning armature to resistor DBR so as to cause dynamic braking. This action does not affect the "Forward" or "Reverse" contactors or change the position of the F or R contacts; the turning armature still is connected to the armature-rectifier terminal 45 and to point 2. When the speed of the motor drops to the desired value, its armature voltage (45-to-2) decreases so as to turn off tube W and drop out QSR and DB just before thyratrons 1 and 2 again pass current to drive the motor.

In detail, if speed dial $1VR$ in Fig. 15N is turned suddenly upward to ask for a lower speed (below base speed), the potential at $1VR$ slider and point 9 becomes more positive; this raises the potential at $W1VR$ slider and grid 5 of tube W, to pick up QSR and DB as described above.† Below tube W, contacts of QSR disconnect capacitor $W3C$ from grid 5 but let $W3C$ discharge through $W6R$; quickly point 6 drops to the negative potential of point 1.

As the motor is braked to a lower speed and its 45-to-2 voltage decreases, the potential at $W1VR$ and grid 5 drops also; soon the

* Below base speed, $1VR$ delays the firing of tubes 1 and 2 until the motor back emf is greater than the anode voltage available from these tubes. Above base speed, $2VR$ strengthens the motor field so that the motor armature generates greater voltage than tubes 1 and 2 can supply. In either case, the motor back emf holds the thyratron cathodes so high that these tubes are phased off by their grids; no current can flow through tubes 1 and 2.

† The purpose of capacitor $W3C$ (at the bottom of Fig. 15N) will be explained later. Before QSR is picked up, $W3C$ is charged to the voltage between tube-W grid 5 and bottom point 1. Although $W3C$ alone may resist a sudden potential rise at grid 5, another capacitor $W2C$ is added to partially offset this effect; $W2C$ and $W3C$ act as a capacitor divider so that, when the 9-to-1 voltage increases suddenly, part of this increase appears across $W2C$ and part of it across $W3C$, permitting the point-5 potential to rise also.

tube-W current stops, so the QSR and DB contacts open. However, the armature voltage 45-to-2 may rise and operate the DB contacts a second time.* It is the purpose of $W3C$ to prevent such further operation of QSR. When QSR drops out after braking the motor, the QSR contact below tube W connects grid 5 to the potential at point 6, which is so negative (as explained above) that tube W is locked out; its grid 5 does not respond to the voltage rise at $W1VR$ caused by a jump of 45-to-2 voltage. While

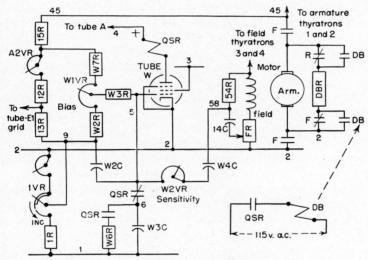

Fig. 15N. A quick slow-down accessory for Thy-mo-trol drive.

$W3C$ is being recharged by electrons flowing through $W3R$, grid 5 remains more negative than $W1VR$ slider; after about $\frac{1}{5}$ sec $W3C$ is nearly charged and tube W again can respond to a potential rise at $W1VR$ slider.

If the motor is running above base speed when the speed dial is turned suddenly to ask for lower speed, $2VR$ (in Fig. 15G) lowers the tube-$G1$ grid and increases the motor-field current. With strengthened field, the turning motor generates greater voltage 45-to-2, raising the $W1VR$-slider potential in Fig. 15N to turn on

* During dynamic braking, the 45-to-2 voltage is equal to the generated voltage E minus the motor's internal voltage drop IR caused by the flow of braking current. When the DB contacts open, this internal IR drop disappears, so the 45-to-2 voltage jumps to the full value E; if tubes 1 and 2 pass current at this same moment, the 45-to-2 voltage jumps further, to the value $E + IR$.

tube W. However, there is also a faster action. As the motor-field voltage increases, capacitor $14C$ forces point 58 positive as is described in Sec. 15–19; through capacitor $W4C$ and sensitivity adjuster $W2VR$, this sudden rise is applied to grid 5, to make tube W pick up QSR and apply dynamic braking to slow the motor.

15–21. Motor-speed Control by Solenoid (CR7507–N115).

Instead of turning a dial to control the motor speed, a small solenoid may be used, with its accessory electronic unit, to furnish the speed signal to the motor-armature circuit of a Thy-mo-trol drive. In Fig. 15O, the armature of a solenoid is moved when a dancer roll is raised or lowered by a continuous strip or web of material passing onto a wind-up roll, driven by motor M; in this way the material is wound under constant tension. As the roll becomes larger, the motor-M speed must be decreased if material is received at nearly constant rate.

If the roll pulls the web faster than it is supplied at A, the web raises the dancer roll, withdrawing the solenoid armature; this increases the current in tube R so that more voltage is produced across $1P$. Part of this $1P$ voltage is used as a speed signal for the Thy-mo-trol drive; it raises the tube-$E1$ grid and decreases the current in BSX so that the motor-M armature receives less voltage. This reduces the motor speed so that the web holds the dancer roll and the solenoid armature at a new position; the whole circuit becomes a closed-cycle system that regulates the solenoid and the motor speed to match the increasing size of the roll.*

Tube R acts as a full-wave rectifier to produce d.c. across $1P$. Although R is a high-vacuum tube, its grid circuits include a phase-shifting network similar to that described for a thyratron in Sec. 14–6 and Fig. 14G.† As the solenoid armature is withdrawn, the a-c voltage across $103R$ becomes more nearly in phase with the anode voltage of left-hand tube R; similarly the voltage across

---

* Note that this speed change is obtained by armature control only. The field-control portion of Fig. 15G is not used when the accessory of Fig. 15O is added.

† This network includes $105R$ and the solenoid coil, connected across the $S1T$ winding, so as to produce out-of-phase a-c voltage across $103R$ and $104R$. The solenoid in Fig. 15O acts as a variable inductance, much like the saturable reactor $1SR$ in Fig. 14G. When the armature is withdrawn, this decreases the solenoid inductance, letting more a.c. flow through its coil; in the same way, increased d.c. in $1SR$ saturates the reactor and decreases the reactor inductance.

104R becomes more nearly in phase with the right-hand tube-R anode voltage. This increases the tube-R current so that more electrons flow from 1T center tap, through 1P and tube R to 1T, producing a pulsing d-c voltage across 1P (filtered by capacitors in the Thy-mo-trol drive, but not shown) that raises terminal 11 more positive than slider 15. As point 11 rises and increases the tube-E1 current, less current flows in tube D1 and the d-c winding of BSX, so that thyratrons 1 and 2 (in Fig. 15G) fire later and apply less voltage to the motor-M armature, as is explained in

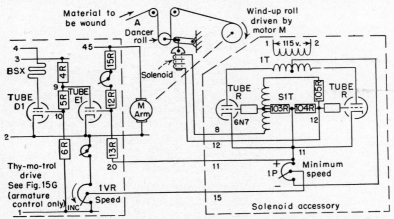

Fig. 15O. Constant-tension control by Thy-mo-trol drive with solenoid accessory (CR7507–N115).

Sec. 15–5. In Fig. 15O the 1VR slider sets the top speed of motor M; even with no voltage across 1P, the tube-E1 grid cannot be lowered below the 1VR-slider potential.

15–22. Motor-speed Control by Selsyn (CR7507–N103). The speed-control dial of the Thy-mo-trol drive (1VR and 2VR in Fig. 15J) may be replaced by a selsyn unit and its electronic accessory, as is shown in Fig. 15P. This selsyn generator* is like a small motor having a three-phase stator winding and a single-phase rotor that may be turned through part of a revolution. When the selsyn rotor is set in one selected position (corresponding to arrow A, at upper right in Fig. 15P), its windings produce no voltage between two of its leads (across transformer 10T), but between another two leads a large voltage is applied to 11T. This position A causes zero speed of the motor controlled by the Thy-

* Chute, "Electronics in Industry," Chap. 28.

mo-trol drive, for the motor receives no armature voltage but receives full field voltage. If the selsyn rotor now is turned slowly clockwise through one-sixth of a turn, or 60 deg, the voltage across $10T$ gradually increases, while the $11T$ voltage remains high; position B provides base speed—full armature and field voltages. When turned beyond B toward C, the selsyn's voltage across $11T$ decreases while the $10T$ voltage remains high. The selsyn may turn through 110 deg, and at C only a small voltage remains across $11T$; this C position gives greatest motor speed, for the motor has full armature voltage and weakened field.

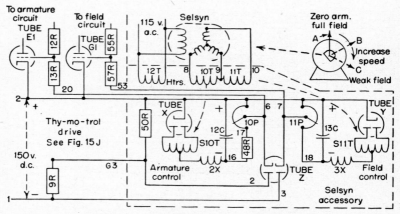

Fig. 15P. Selsyn accessory (CR7507–N103) for Thy-mo-trol drive.

In the selsyn-accessory circuit in Fig. 15P, any voltage applied to transformer $10T$ produces a similar voltage at $S10T$ that is rectified by tube X; electrons flow from $S10T$ through $2X$, $48R$, $10P$ and tube X, charging capacitor $12C$. As the selsyn is turned, applying a varying voltage to $10T$, the d-c voltage across $12C$ changes gradually, as is shown in Fig. 15Q. Similarly, any voltage applied through transformer $11T$ is rectified by tube Y so as to charge capacitor $13C$.

In Fig. 15P, the $13C$ voltage appears also across $11P$, whose slider 7 connects to the motor-field-control circuit of the Thy-mo-trol drive at point 53 and the grid of tube $G1$. However, with $11P$ turned clockwise to touch at 18, tube Z may limit the amount of voltage across $11P$. Anode 3 of tube Z connects to point 1 of the Thy-mo-trol circuit, which is steadily 150 volts below (more negative than) point 2 or the top side of $11P$. So long as $11T$

produces less than 150 volts across 13C, tube Z has no effect on the potential at 11P slider 7. However, if the 13C voltage tries to exceed 150 volts, the tube-Z cathode 7 is forced below anode 3; tube Z "takes over," for current flows through it to prevent point 7 from becoming more negative than point 1. At D in Fig. 15Q, transformer 11T tries to produce more than 150 volts across 13C, but slider 7 is held at this −150-volt potential by the take-over action of tube Z. So, although 11T receives a varying voltage from the selsyn as it is turned from A to B, this voltage remains so large that tube Z holds slider 7 at the constant potential of point 1.

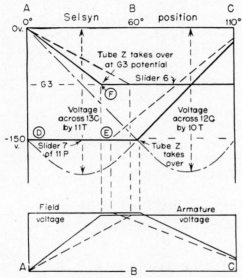

FIG. 15Q. Voltages produced as selsyn is turned in Fig. 15P.

Inasmuch as the tube-$G1$ grid is held steadily at low potential, this permits maximum current through tube $D2$ and CSX in Fig. 15J so that thyratrons 3 and 4 provide full current for the motor field. Between selsyn positions B and C, Fig. 15Q shows how the 13C voltage becomes less than 150 volts; here cathode 7 of tube Z becomes more positive than its anode 3 so tube Z disconnects slider 7 from the −150-volt potential. As the selsyn is turned further toward C, the 13C voltage becomes less and raises the tube-$G1$ grid, to decrease the motor-field current and to further raise the motor speed.

If the 11P slider now is turned (ccw) so that only a part of the

13C voltage reaches slider 7, cathode 7 of tube Z may rise above anode 3, following the dashed line at point E, before the selsyn has reached the B position. This permits the motor-field current to decrease before the motor-armature voltage has risen to its full amount, providing an overlap adjustment.

A similar action occurs in the motor-armature-control circuit as the selsyn is turned from A toward B. Notice that the tube-Z anode 2 is connected to point $G3$ on a voltage divider (9R–50R) that holds anode 2 about 75 volts below point 2 of the Thy-mo-trol drive. As the voltage across 12C increases, a part of this voltage gradually lowers the potential of the 10P slider 6, thereby lowering point 20 and the tube-$E1$ grid, so that the motor-armature voltage may increase. Full armature voltage results when point 20 is about 75 volts below point 2, as at F. At this voltage, cathode 6 of tube Z has dropped to its anode-2 potential, so current flows through tube Z to hold the 10P slider 6 at this -75-volt potential, although the selsyn applies greater voltages through 10T when turned between B and C.

If 10P now is turned so that less of the 12C voltage appears at slider 6, the selsyn may reach its position B before the point-6 potential (following the dashed line in Fig. 15P) reaches the -75-volt value and provides full armature voltage.

CHAPTER 16

MOTOR–GENERATOR CONTROL OF MOTOR SPEED

Although an all-tube equipment may be used for variable-speed motors of 10 to 1000 hp, it may be desirable to supply these large d-c motors from separate motor-generator sets, as is described in Sec. 13–1 and Fig. 13B. A single-package type of such a drive (consisting of the separately mounted variable-speed motor, its m-g set and all controls) is the speed variator; its simpler form, having no electronic controls, is sketched in Fig. 16A. To obtain closer speed control and other features (similar to the Thy-mo-trol drive described in Chap. 15), the fields of the generator and the d-c motor may be supplied by electronic circuits; such an electronic speed variator is shown in Fig. 16C and is explained later.

16–1. M-g Set for Speed Control. To show quickly how the speed of a d-c motor is varied by controlling a motor-generator set, Fig. 16A shows such a motor M receiving armature voltage from generator G. The shunt fields of G and M receive d.c. from a small generator or exciter E. Both E and G are driven at constant speed by the a-c motor K, forming the m-g set. At the upper left in Fig. 16A, the start button picks up contactor AM, bringing the m-g set up to speed. Exciter E builds up its own voltage and supplies field current to motor M; generator G produces no armature voltage, since the open DM contact keeps voltage away from the G shunt field. When the lower start button is closed, the exciter voltage may pick up d-c contactor DM; the main DM contact connects the M armature to the G armature, while another DM contact applies d-c voltage across generator field rheostat $1Rh$. If the speed dial is turned (counterclockwise) to the zero position, none of the $1Rh$ voltage reaches the generator field and G produces so little voltage that motor M hardly turns. This condition is shown at the bottom of Fig. 16B; with no generator-field current the speed of motor M may be zero, although the motor-M field is at 100 per cent strength. Both $1Rh$ and $2Rh$ are turned by the speed dial; if they are turned clockwise gradually, to raise the M speed, the slider moves along the resistance section of 1Rh, increasing the voltage across the G shunt field. This makes G produce more armature voltage which, applied to the M armature, raises the motor speed. When the speed dial has been turned

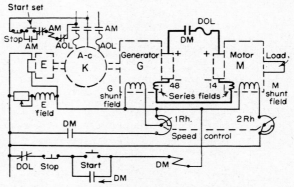

FIG. 16A. Circuit of simple speed variator.

through one-third of its travel, the 1Rh slider has moved across two-thirds of its resistance; with 67 per cent of G field current, Fig. 16B shows the motor running at about two-thirds of its base speed. Meanwhile motor M has 100 per cent field current. Above base speed, the generator receives its full field current so that full voltage is applied to the motor-M armature. Turning beyond mid-point has no further effect at 1Rh, but the 2Rh slider now inserts resistance in series with the motor-M field. With full armature voltage, the motor-M speed rises as the M-field current is decreased, as is shown in the upper part of Fig. 16B.

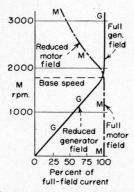

FIG. 16B. Motor speed vs. strength of motor field or generator field.

While contact DM in Fig. 16A connects the M armature to the G armature, power may flow from G to M so long as the G voltage is greater than the M voltage. Notice that the upper side of both armatures (connected together by DM) is more positive than the lower armature terminals 48 and 14; motor M, while turning, generates a back voltage nearly as large as the G voltage. While the motor-M speed is rising, or while G furnishes power to M to drive its load, the G armature voltage is slightly greater than the

M voltage, so G terminal 48 is slightly below (more negative than) M terminal 14. This 48-to-14 voltage appears across the G and M series fields; the amount of this voltage increases as the motor draws more armature (load) current. (This 48-to-14 voltage is used in Sec. 16–10 as a motor-load signal to the electronic circuits.) However, when motor M is being slowed, as by turning the speed dial to a lower setting, the M back voltage becomes greater than the G voltage (either because the M field is strengthened, or because the G field is weakened). Motor M here "pumps back" power to G (and through motor K into the a-c supply line); point 48 now is more positive than 14, and this reversed 48-to-14 voltage increases as M sends more power back to G.

In the basic system described above, the speed of motor M may change as much as 5 or 10 per cent when the motor load changes; this is natural in most variable-voltage (or Ward-Leonard) d-c systems unless circuits are added to offset these changes.

16–2. Electronic Speed Variators. To provide close speed control, load-limit protection and similar advantages, an electronic control unit is added to the speed-variator system described above. This equipment is shown in Fig. 16C, not including the variable-speed motor M; the front of the control panel appears in Fig. 16D, and the system diagram is shown in Fig. 16F. In place of the exciter E (of Fig. 16A), this electronic variator uses thyratrons 1 and 2 to furnish direct current to the generator-G field; thyratrons 3 and 4 similarly supply the field of motor M.

At the upper right in Fig. 16F, generator G and motor M are arranged in an armature loop circuit closed by contact DM (as in Fig. 16A).* To vary the motor-M speed, the currents in the G shunt field and the M shunt field are changed as is outlined in Sec. 16–1 and Fig. 16B; these changes are produced by phase-shift control of thyratrons 1, 2, 3 and 4. These thyratron circuits are merely outlined in Fig. 16F; tubes 3 and 4 control the amount of current in the motor-M field just like tubes 3 and 4 shown in Figs. 15G or 15I, while tubes 1 and 2 control the generator-G field current in the same manner. The details of the phase-shifting bridge (such as saturable reactor MSX, $6R$ and $P2T$, at upper

---

* Other relays, not shown in Fig. 16F, prevent the closing of DM unless generator G is supplying less than 40 volts. Later, when the stop button drops out most relays, one relay contact closes to connect the $1P$ slider to center line 14, to decrease the generator field current; the DM contacts are held closed until the motor or generator armature voltage becomes less than 20 volts.

264   ELECTRONIC MOTOR AND WELDER CONTROLS

Fig. 16C. Speed variator with electronic control (rear view).

Fig. 16D. Electronic control panel for speed variator.

MOTOR–GENERATOR CONTROL OF MOTOR SPEED

right in Fig. 16F) are explained in Sec. 13–6; the grid-circuit wave shapes are discussed in Sec. 15–18 and Fig. 15M.

At the upper left in Fig. 16F, the d-c winding of saturable reactor MSX receives current through tube $F2$. Whenever the grid potential at 34 rises, increasing the tube-$F2$ current, MSX becomes more saturated and the a-c voltage decreases across the MSX windings (at upper right), advancing the firing point of thyratrons 3 and 4 to increase the M field current. Briefly, the motor-M field current increases as tube-$F2$ current increases. Similarly, the generator-G field current increases as tube-$D2$ current increases through GSX. Most circuits in Fig. 16F serve to control tubes $D2$ and $F2$.

Two forms of electronic speed variator will be described. The first form provides timed acceleration (that brings motor M up to speed at a selected time rate or similarly slows it down); the second form provides current-limit control (that brings motor M up to speed without exceeding a selected value of armature current) that permits only a selected amount of motor overload at any time. Both forms use all circuits shown in Fig. 16F, except those at the lower right. Whereas the first form (Sec. 16–3 and Fig. 16F) uses tubes $F1$, G and H in a time-delay circuit, the second form (Sec. 16–10 and Fig. 16H) uses tubes $F1$, G and H in a current-limit circuit.

16–3. Timed-acceleration Variator Control (CR7513–E101).

This first form of electronic speed variator (as described above) is diagramed in Fig. 16F. The entire speed range of motor M is controlled by the single dial $1P$ (at the right). First let us connect the $1P$ slider to point 42, cathode of tube $F1$; this removes the timed-acceleration circuit (to be explained in Sec. 16–9), and lets us explain the rest of Fig. 16F as it is used in both forms of electronic variator.

Briefly, when the $1P$ slider (set for zero speed) and point 42 are at the high potential of center line 14, tube $D1$ is turned on, so that tube $D2$ passes no current through reactor GSX; thyratrons 1 and 2 are phased off, so there is no current in the generator-G field nor is there any G armature voltage. At the same time, the high potential of 42 turns on tubes $E1$ and $E2$ (at lower left in Fig. 16F), raising their cathodes and grid 34 so that tube $F2$ passes current through reactor MSX; thyratrons 3 and 4 supply full field current to motor M. Now as $1P$ is turned clockwise to ask for more speed, the potential of point 42 is lowered so that the tube-$D1$ current decreases, letting tube $D2$ pass current through GSX; tubes 1 and 2 fire earlier, increasing the G field current and the G voltage applied to the armature of motor M. This greater G voltage raises the G terminal 49, bringing the grid of tube $D1$ back to the point of balance. As the $1P$ slider reaches the base-speed position, rectifier 3Rec disconnects point 42 from 43, preventing further effect on tube $D1$ or the generator-G voltage; however, point 42 now lowers the grids of tubes $E1$ and $E2$ so that their cathode potentials drop also, decreasing the current through tube $F2$ and MSX, so that thyratrons 3 and 4 weaken the motor-M field to permit speed above the motor's base speed. Meanwhile, if motor M becomes heavily loaded, the increased armature current drives G terminal 48 more negative than 14; $8P$ uses part of this 48-to-14 voltage to decrease the tube-$D1$ current, turning on tubes $D2$, 1 and 2 to provide greater armature voltage to maintain the motor speed despite changes of load.

In detail, tube A (upper left in Fig. 16F) rectifies the a-c supply voltage so that d.c. (filtered by $1X$ and $1C$) appears between top line 15 and bottom line 12. The voltage-regulator tubes B and C hold line 15 about 100 volts above center line 14; 14 is about 100 volts above 13. The excess voltage from tube A appears across the buffer resistor $15R$, so that line 12 is more negative than 13 (see Sec. 2–4, footnote).

16–4. Control of Motor Field. At the left in Fig. 16F, tubes $E1$, $E2$ and $F2$ control the motor-M field current. As is explained in Sec. 13–2, if current decreases in tube $F2$, the field current of motor M also decreases, raising the motor speed.

With a jumper connecting the $1P$ slider to cathode 42 of tube $F1$ (as added in Sec. 16–3), let us see how $1P$ controls the motor-field current. This circuit appears also in Fig. 16E. Set for low speed, the $1P$ slider and point 42 are close to the potential of line 14. However, owing to resistor $24R$, we shall see that grid 41 of tubes $E1$ and $E2$ is at a potential lower than point 42.

Any current through tube $E2$ passes also through its cathode resistor $35R$ and anode resistors $32R$ and $9P$. If all the $9P$ resistance is in the circuit,* there is enough tube-$E2$ current so that

* Overlap adjuster $9P$ sets the point (on the $1P$ speed dial) where the motor-M field current starts to decrease. With $9P$ shorted (turned clockwise, for greatest overlap), the field is weakened before the motor-armature voltage rises above 180 volts (for a 250-volt motor). With all the $9P$ resistance in circuit, the motor armature receives nearly 240 volts before the motor field current decreases. See further Sec. 16–6.

cathode 18 of tube $E2$ is about 70 volts above (more positive than) line 13. Grid 41 is very near this cathode-18 potential, since electrons flow from cathode 18 to grid 41 and through $24R$ to 42. So, while $1P$ is being turned downward through its low-speed range, the point-42 potential gradually is lowered from $+100$ volts to $+70$ volts above line 13 (as is shown at A in Fig. 16E); this has no effect on grids 41, for they remain about 70 volts above 13, as shown at B. Cathodes 17 and 18 of tubes $E1$ and

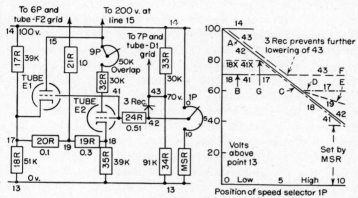

Fig. 16E. Motor-field control in circuits shown in Fig. 16F and Fig. 16H.

$E2$ also are near $+70$-volt potential; this provides high potential at point 19 on the voltage divider $23R$–$6P$–$21R$–$20R$ and $19R$, so that grid 34 of tube $F2$ permits enough tube-$F2$ current to let motor M receive full field current.*

16–5. Motor-field-shaping Circuit.† When $1P$ is turned downward in Fig. 16E, to obtain higher speed, point 42 may become less than 70 volts above line 13. Grid 41 now becomes slightly more negative than the tube-$E2$ cathode; tube-$E2$ grid current no

---

* Motor-field-current adjuster $5P$ may reduce this maximum current by being turned clockwise to a lower potential, thereby decreasing the tube-$F2$ current. Then, with $1P$ set for top speed, $6P$ may adjust the minimum value of tube-$F2$ current and thereby limit this top speed to the desired safe value.

† In Fig. 16B the reduced-motor-field curve shows that, unless the field current is lowered at a gradually decreasing rate, the motor speed will rise a different amount for each 5 per cent change in speed-dial setting. Therefore a field-shaping circuit is added so that, at high motor speeds, the motor-field current is decreased more slowly than at medium speeds. During timed acceleration (in Sec. 16–9), the speed-signal potential at point 42 drops at a uniform rate; this produces too rapid acceleration near top speed unless the field-shaping circuit is added.

longer flows through $24R$, so the point-41 potential decreases along with the point-42 potential. As grids 41 are lowered, current decreases in both tubes $E1$ and $E2$; both cathodes 17 and 18 are lowered, drawing point 19 downward at the same rate (as at C in Fig. 16E) and thereby decreasing the tube-$F2$ current and the motor-M field current.*

Notice here that cathode 17 of tube $E1$ is connected to voltage divider $17R$–$18R$; these resistors hold point 17 at $+60$ volts (above 13) unless tube $E1$ passes current through $18R$. When grid 41 drops below $+60$ volts, the tube-$E1$ current stops and cathode 17 cannot follow the grid downward; point 17 now remains at constant potential (at D in Fig. 16E), while cathode 18 continues to follow grid 41 downward as the 1P-slider potential is turned clockwise to lower potential. Here resistors $19R$ and $20R$ hold point 19 part way between 17 and 18; since 17 remains steady while 18 continues to drop, the point-19 potential drops more slowly than before, as is shown at E. Thus the field current of motor M continues to decrease, but at a slower rate, and provides more gradual control as motor M approaches top speed.

While the 1P slider and point-42 potentials are below $+70$ volts, point 42 is more negative than point 43 (on voltage divider $33R$–$34R$), so selenium rectifier 3Rec disconnects 42 from 43; the point-43 potential does not follow point 42 downward below the $+70$-volt level, as is shown at F in Fig. 16E. So, while motor M is running above base speed, the turning of 1P may further weaken the motor field without affecting the generator-G field and the motor-M armature voltage.

16–6. Control of Generator Field and Motor-armature Voltage. Returning to Fig. 16F, let us use 1P to control motor M at low speed, by varying the field current of generator G. (Retain the jumper, added in Sec. 16–3, that connects the 1P slider to cathode 42 of tube $F1$. Also let the start button close contact DM, connecting G to M.) When set for bottom speed, the 1P slider and point 42 are at the same potential as center line 14. Since current

---

* Below tube $F2$, in Fig. 16F, cathode resistor $12R$ provides more stable circuit operation as is mentioned in Sec. 18–13. Resistor $11R$ has a similar effect below tube $D2$, to stabilize the G field circuit. Resistor $12R$ is connected in the motor-M field circuit, as shown at the right in Fig. 16F. When a signal raises the tube-$F2$ grid 34 by 3 volts, the M field current increases, causing greater voltage drop across $12R$; this raises cathode 23 by perhaps 2 volts, so that the net increase in tube-$F2$ grid voltage is only 1 volt (grid to cathode).

now may pass through rectifier 3Rec, point 43 is raised to the high potential of point 42; these electrons flow from line 13 through 34R to 43, through 3Rec to 42, and by jumper to the 1P slider, which has been turned to touch at 14. Under these conditions, we shall see that generator G produces no voltage or current, so the G terminal 49 is at the same potential as G terminal 48; both 49 and 48 are at line-14 potential. Connected between points 49 and 14 is the voltage divider 38R–39R–8P (such that the 46-to-14 voltage is about one-fifth of the G-armature voltage 49-to-14); between 46 and 43 is the divider 37R–7P–36R. Since both 49 and 43 are at line-14 potential, all parts of these dividers are at 14 potential; this includes grid 37 of tube $D1$. Cathode 31 of tube $D1$ is raised a little above 14, because of the bias setting of 4P (if 3P is shorted, for greatest sensitivity); so, with grid 37 at 14 potential, there is enough current flowing through tube $D1$ and 26R to lower the anode-29 potential by about 70 volts. Connected between 29 and 13 is a voltage divider 27R–28R, which now holds grid 28 of tube $D2$ at a potential below point 14 so that tube $D2$ passes no current through the d-c winding of saturable reactor GSX. Thyratrons 1 and 2 are turned off; generator G receives no field current and produces no voltage for the motor-M armature.

Now let us turn 1P clockwise to raise the motor-M speed. This lowers point 42, the 7P slider and grid 37, so that current decreases through tube $D1$ and 26R; the rise of point 29 and grid 28 lets tube $D2$ pass current through GSX, so generator G supplies armature voltage to speed up the motor M.* While point 42 is being moved downward, furnishing the "speed reference" voltage 43–14 that turns off tube $D1$, the generator-G voltage is increasing, raising its terminal 49 and supplying a "feed-back" signal 46–14 to match or oppose the 43–14 reference. If 1P and 43 are lowered so that grid 37 is lowered 5 volts, the generator G voltage is thereby increased, raising points 49 and 46 enough to return grid 37 upward about 4½ volts; only the ½-volt difference is needed to turn off tube $D1$ enough to cause the desired G field current and motor-M speed.

* The effect of 11R below tube $D2$ is mentioned in footnote opposite. To prevent the G armature voltage from increasing too quickly or causing too much armature current, the timed acceleration circuit is added between the 1P slider and point 42, as is described in Sec. 16–9; or, instead, a current-limit circuit may be added, as in Sec. 16–10.

When $1P$ is brought to a medium-speed position, such that points 42 and 43 are about 30 volts below 14, Fig. 16E shows that 42 now has dropped to the potential set by voltage divider $33R$–$34R$. So long as 42 is more than 70 volts above 13, rectifier 3Rec passes current to raise point 43 along with point 42. However, when 42 drops below this 70-volt point, the rectifying (one-way) action of 3Rec disconnects 42 from 43; further lowering of point 42 has no effect on 43, which remains at the potential set by $33R$–$34R$. At this final potential of point 43, generator G furnishes full armature voltage to motor M. Generator G may produce perhaps 180 volts (49-to-48) or enough to raise point 46 so as to bring grid 37 to a potential slightly below the tube $D1$ cathode. If $7P$ now is turned clockwise, lowering the tube-$D1$ grid 37, G must generate greater voltage to restore the $7P$ and grid-37 potentials to balance; in this way $7P$ may adjust the maximum voltage (such as 250 volts) that G applies to motor M while the motor is lightly loaded.

As point 42 drops near to the 70-volt (above 13) potential that provides full armature voltage, the shunt field of motor M is being weakened as is described in Sec. 16–5. If $9P$ is set for small overlap (as mentioned in Sec. 16–4, footnote), neither the tube-$F2$ current nor the motor-M field current starts to decrease until motor M receives nearly full armature voltage, as indicated at B in Fig. 16E. However, when $9P$ is turned clockwise for large overlap, grid 41 is held at higher potential; $5P$ adjusts the full-field current at this setting. At G in Fig. 16E we see that the tube-$E2$ grid (here shown as $41X$) starts to lose potential, to weaken the motor field, before the G field and armature voltage approach their full values.

16–7. Sensitivity Control. At the center of Fig. 16F, the cathode circuit of tube $D1$ includes adjusters $2P$, $3P$ and $4P$. If $3P$ is shorted, cathode 31 is at a steady potential at the $4P$ slider and is not affected by the circuit through $2P$; a signal at grid 37 produces maximum change in tube-$D1$ current and generator-G voltage. However, as $3P$ is turned clockwise and inserts resistance, part of the tube-$D1$ anode current now flows through $2P$ and $29R$. As the tube-$D1$ anode current increases, caused by a 3-volt rise at grid 37, this current increases the voltage drop across $3P$, also the drop across $2P$ and $29R$, so that cathode 31 rises perhaps 2 volts. The 3-volt rise at 37 has caused only a 1-volt (grid-to-cathode) change in grid voltage; a given signal causes

MOTOR–GENERATOR CONTROL OF MOTOR SPEED 271

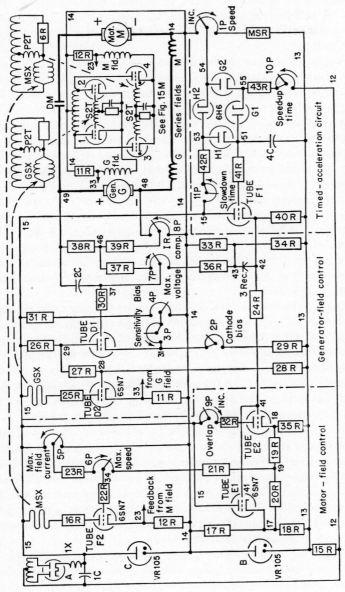

Fig. 16F. Circuit of timed-acceleration variator control (CR7513–E101).

less change in tube-$D1$ current or generator-G voltage, so the circuit sensitivity is thus decreased. Because of this cathode-follower[13-5] action of tube $D1$, the circuit becomes more stable (less likely to hunt).

16–8. Speed Correction for Motor-load Changes. When a d-c motor receives greater load at its shaft, it may tend to slow down (because of its internal "IR drop" or voltage loss), even though it receives constant armature voltage. Therefore, to hold steady the motor-M speed when the load increases, generator G must be made to produce greater voltage. In Fig. 16F, 8P may be set as a speed-drop adjuster, also called IR-drop compensation. As motor M becomes more heavily loaded, more current flows from G through the M armature and the series (and commutating) fields; the voltage drop across these fields drives point 48 more negative than 14. This lowers the potential at the 8P slider, thereby lowering the potential of point 46 on divider $38R$–$39R$–$8P$; since this lowers grid 37 of tube $D1$, tube $D2$ is made to increase the field strength of generator G so that G supplies enough more voltage to motor M to offset the motor's increased IR drop, and thereby to hold constant the motor speed.

Since the potential at 8P also tends to raise the generator voltage while the motor is starting and tends to increase motor current during both starting and slow-down, there is greater need that a circuit be added to control the motor's speed-up or slow-down action.

16–9. Timed Speed-up and Slow-down. The connection (from 1P slider to cathode 42 of tube $F1$, as added in Sec. 16–3) should now be removed at the lower right in Fig. 16F, to let tubes $F1$, G and H control motor M while its speed is being changed.

Notice that tube $F1$ acts as a cathode follower; the tube load resistor $40R$ is in the tube's cathode circuit, while the tube anode is connected to the high constant potential at point 15. With this connection, the potential of cathode 42 follows the tube-$F1$ grid potential; if grid 51 drops 5 volts, cathode 42 drops about 5 volts; if 51 is at center line-14 potential, cathode 42 is close to 14 potential.

Tubes $G1$, $G2$, $H1$ and $H2$ act as simple rectifiers, to "take over" the control. Let us watch their action.

While set for bottom motor speed, 1P holds point 54 at center line-14 potential; therefore point 53 also is near 14 potential, for electrons flow from 54 through tube $H2$ to 53, and through $42R$ and 11P to 15; there is very little voltage across tube $H2$ while current flows through it. Similarly, point 51 is close to 14 potential, since electrons may flow from bottom point 12 through 10P, 43R, tube $G1$ and tube $H1$ to 53. Capacitor 4C is charged to the voltage between points 14 and 13, as indicated in Fig. 16G. Cathode-follower tube $F1$ holds point 42 close to 14 potential.

Suppose that the speed dial 1P is quickly turned through its entire range, to ask for top speed; the potential of point 54 drops suddenly to lower potential, in Fig. 16G. This produces no sudden

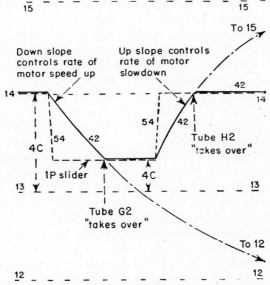

Fig. 16G. Voltages in Fig. 16F to control speed-up and slow-down.

change at point 42, since the voltage across 4C cannot change suddenly. Instead, the lowering of point 54 lowers point 53 also, so that the anodes of tubes $H1$ and $G2$ are more negative than their cathodes 51 and 55; current stops flowing in tubes $H1$ and $G2$. Point 51 remains connected through tube $G1$ to point 55; electrons flow from 12 through 10P, 43R and tube $G1$ so as to lower the potential of point 51, discharging 4C at the steady rate set by the amount of resistance in adjuster 10P. Therefore tube $F1$ lowers the point-42 potential at this same steady rate.* Although 1P is

* Figure 16G shows that the point-42 potential drops along a steady slope, since 4C is discharging along a curve that finally would bring point 51 to the negative potential of point 12. Similarly, when 4C is charged through tube $H1$ during slow-down time, point 51 rises along a steady slope that curves later as it approaches the high point-15 potential.

turned suddenly, the motor-M speed rises steadily during the speed-up time selected by $10P$. When $4C$ has discharged so that point 51 drops to the potential of the $1P$ slider 54, tube $G2$ "takes over," passing current to hold 51 and 55 at this 54 potential; tube $F1$ holds 42 at this steady lower potential (shown in Fig. 16G) that causes top motor speed, as explained in Sec. 16–3.

If $1P$ now is turned suddenly to a lower-speed position, thereby suddenly raising the point-54 potential, point 42 does not follow at this same rate. Instead, the raised potential of 54 makes the cathodes of tubes $H2$ and $G1$ (through $G2$) rise higher than their anodes; tubes $H2$ and $G1$ disconnect points 51 and 53 from points 54 and 55. Capacitor $4C$ charges to higher voltage as electrons flow from its terminal 51 through tube $H1$, $42R$ and $11P$ to positive point 15; $11P$ sets the time required for tube $F1$ to raise point 42 and thereby to slow the motor. When point 51 rises to the potential of the $1P$ slider 54, tube $H2$ "takes over" by passing current that connects points 53 and 54.

16–10. Electronic Variator Control with Current Limit (CR7513–E118). Instead of raising or lowering the motor speed at a timed rate, as is explained above, the variator may have a current-limit circuit that prevents motor M from drawing more than a desired amount of armature current. Figure 16H shows how this current-limit circuit is added between the motor-field control and the generator-field control of Fig. 16F. Tubes $F1$, G and H (used previously in the timed-acceleration circuit) now act when too much current tries to flow in the motor armature; these tubes control the motor-field and generator-field circuits so as to limit this flow of armature current, thereby preventing motor overload or too rapid speed-up or slow-down.

Briefly, when the d-c motor M draws current, while starting or running as a motor, the potential of point 48 (at upper right in Fig. 16H) becomes more negative than center line 14. When this signal at 48 becomes stronger than the value selected by $10P$, the current decreases through tube $F1$ and $40R$; in the right sequence, tube $H2$ acts on tube $F2$ to strengthen the motor-M field, then tube $G2$ acts on tube $D1$ to weaken the generator-G field and decrease the generator's armature voltage. Both of these actions make motor M draw less load current. When motor M is being slowed, current reverses in the series fields so that point 48 becomes more positive than line 14 (see Sec. 16–1); when 48 rises above the value selected by $11P$, current increases through tube $F1$ and $40R$; first, tube $H1$ acts on tube $D1$ to raise the armature voltage of generator G, then tube $G1$ acts on tube $F2$ to weaken the motor-M field. These actions decrease the current that motor M can "pump back" into G.

Before motor M is started, no current flows through the G or M series fields, so there is no voltage across $8P$; the $8P$ terminal 48 is at the potential of center line 14. Point 48 connects to the middle of voltage divider $41R$–$10P$–$11P$–$43R$ (center of Fig. 16H). In its usual setting, the $10P$ slider 53 now is more positive than line 14 and tries to raise grid 52 of tube $F1$, but is disconnected by the metallic rectifier 4Rec; similarly, the $11P$ slider 51 is more negative than 14 and tries to lower the tube-$F1$ grid, but is disconnected by 5Rec. Since grid 52 is not connected through either 4Rec or 5Rec, 52 is at 14 potential; $12P$ is adjusted to raise cathode 56 until tube $F1$ passes only enough current through $40R$ so that its anode 55 is about 50 volts above line 14. Between 55 and bottom line 13 is the voltage divider $45R$–$46R$–$47R$–$48R$–$49R$; before motor M is started, terminal 59 of $47R$ is about 3 volts above line 14, while terminal 60 is about 3 volts below line 14. Cathode 58 of tube $G1$ is more positive than its anode 34, and anode 60 of tube $H2$ is more negative than its cathode 34, so neither tube passes current; disconnected by these tubes, grid 34 of tube $F2$ is responding only to the motor-field control (through tubes $E1$ and $E2$, as is described in Sec. 16–4). Similarly, cathode 59 of tube $H1$ is slightly above its anode-37 potential, and anode 61 of tube $G2$ is below 37; disconnected by these tubes, grid 37 of tube $D1$ responds only to the generator-field control (through $7P$, $36R$ and 3Rec to 42, as is described in Sec. 16–6).

If motor M is started and brought up to speed by the slow turning of speed dial $1P$, the motor may not draw enough armature current to disturb or change the conditions just described. However, if $1P$ is turned abruptly to ask for higher speed, or if motor M has too much load, the armature current causes enough voltage across $8P$ to drive 48 so far negative that slider 53 of $10P$ falls below line 14; rectifier 4Rec now conducts, letting point 53 pull the tube-$F1$ grid more negative. (Current-limit adjuster $10P$ sets the value of motor-M current that can cause this action.) As the current now decreases in tube $F1$, its anode-55 potential rises, thereby raising points 58, 59, 60 and 61. This has no effect on tubes $G1$ or $H1$. However, when anode 60 has been raised 4 or 5 volts, tube $H2$ may pass current; if 60 rises yet higher, point 34

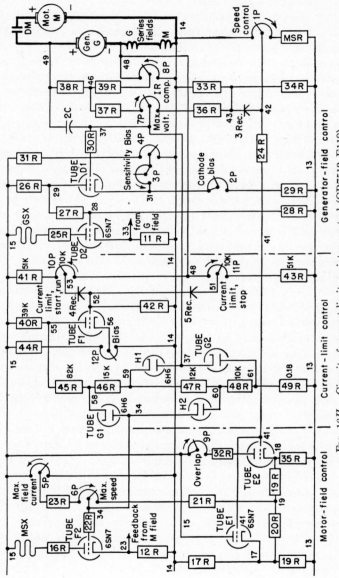

Fig. 16H. Circuit of current-limit variator control (CR7513-E118).

MOTOR–GENERATOR CONTROL OF MOTOR SPEED

is pulled upward also (since tube $H2$ now connects 34 to 60), thereby raising the grid of tube $F2$, to strengthen the motor-M field, as is explained in Sec. 16–2. With further rise at the tube-$F1$ anode 55, point 61 may be raised until the tube-$G2$ anode becomes more positive than its cathode 37; here tube $G2$ "takes over" so that 37 is raised along with 61. This rise of grid-37 potential increases the tube-$D1$ current, decreasing the current through tube $D2$ and GSX; this decreases the generator-G field current and armature voltage, as is explained in Sec. 16–6.

16–11. Slow-down with Current Limit. When $1P$ is turned quickly to ask for lower speed or when the stop button is pushed, letting a relay (not shown) connect point 42 to point 14, the motor-field control in Fig. 16H increases the motor-M field current, so that M generates greater back voltage. At the same time, the generator-field control reduces the generator-G field current, so that G generates less voltage. Motor M may try to "pump back" too much current; flowing through the series fields, this current raises the point-48 potential, until slider 51 of $11P$ reaches a higher potential than center line 14. At this slow-down current limit as set by $11P$, current flows through 5Rec to connect the tube-$F1$ grid 52 to point 51; a further rise at 48 and 51 causes greater current to flow through tube $F1$ and $40R$, lowering point 55. As all parts of voltage divider $45R$–$46R$–$47R$–$48R$ drop to lower potentials, this has no effect on tubes $G2$ or $H2$. However, when cathode 59 of tube $H1$ is lowered 3 or 4 volts, it becomes lower than its anode-37 potential, so tube $H1$ passes current; further lowering of point 59 now pulls grid 37 downward so that the tube-$D1$ current decreases, letting tube $D2$ pass more current through GSX, to increase the generator-G armature voltage so that it may oppose the M back voltage and decrease the pump-back current.

Further lowering of point 55 brings cathode 58 of tube $G1$ below the potential of its anode 34; tube-$F2$ grid 34 then follows downward, decreasing the current through $F2$ and MSX, to decrease the motor-M field current and back voltage, to lessen the pump-back current.

CHAPTER 17

THE ELECTRONIC AMPLIDYNE

In the preceding chapter, thyratron tubes are used to control directly the field currents of large generators and motors. In contrast, a small amplidyne d-c generator here is combined with electron tubes to provide a variable-voltage supply for a d-c motor, so that the motor's speed remains constant at any desired level, or so that the motor receives constant current and produces a desired steady amount of force or tension.

17-1. Electronic Amplidyne (5AM79AB334). This is a general-purpose equipment, shown in Fig. 17A, including a 1½-kw amplidyne generator (see Sec. 14-4, footnote) driven at constant speed by an a-c motor; this 1½-kw generator may supply the power to drive a 1½-hp d-c motor, or it may be used to furnish the field current of a larger d-c generator or motor. As a part of the amplidyne m-g set, a small exciter provides direct current for the shunt field of the 1½-hp d-c motor, and also supplies d-c power to the electronic circuits in the cabinet mounted on the amplidyne set. This packaged equipment combines these electronic amplifier circuits with the high-speed amplifying action of the rotating amplidyne, and generally is used as part of a closed-cycle regulating system.* [12-5]

The circuit of this electronic amplidyne is shown in Fig. 17B. At the right, the amplidyne generator A supplies d.c. to the motor M, in the same way as does generator G in Fig. 16A and Fig. 16F. In Fig. 17B, three jumpers, X, Y and Z, are arranged for "armature feedback"; the motor's armature voltage becomes the speed signal, so that amplidyne A furnishes the right voltage to hold constant the speed of motor M. (Later, jumpers X, Y and Z are rearranged for "tachometer feedback" in Fig. 17C, or for constant-current regulation in Fig. 17D.)

At the upper right in Fig. 17B, a-c power is supplied to motor K

---
* Dutcher, J. L., Torque and Speed Regulation with the Electronic Amplidyne, *Electrical Manufacturing*, July, 1948.

to drive the amplidyne A and exciter E. Through transformer $1T$, separate voltages heat the electron tubes and supply power to tube 1. Tube 1 supplies a rectified d-c voltage that is filtered by $12C$, $29R$ and $11C$ (see Sec. 2–4, footnote); because of voltage-regulator tube 2 and its buffer resistor $28R$, a steady d-c voltage is held across speed-adjuster P, which selects the desired speed of d-c motor M. The voltage between point 14 and slider 13 of P is the "reference voltage" that controls the other tubes and selects the voltage produced by amplidyne A.

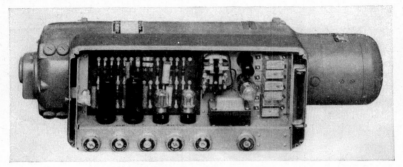

FIG. 17A. The electronic amplidyne.

Briefly, when P is turned downward in Fig. 17B so as to raise the speed of motor M, the slider 13 becomes more negative than 14; as the tube-$3B$ current decreases, raising point 28, grid 32 of tube 5 rises also, so tube 5 strengthens (increases the amount of current through) the R amplidyne field. At the same time, since tubes $3A$ and $3B$ (with $9R$) act as a "long-tailed pair,"[12-8] current increases in tube $3A$, lowering point 29 and grid 33 so as to decrease the tube-6 current and weaken the L amplidyne field. Since the R field is now stronger than the L field, amplidyne A produces armature voltage that is more positive at the A terminal 5, to turn motor M in the forward direction. If motor M tries to draw too much current from A, the voltage produced across the C fields* drives negative points 6 and 17, turning on tube $4B$ to lower the point-28 potential; this opposes the action of tube $3B$ and limits the tube-5 current and the strength of the R field.

At all times, exciter E supplies 230 volts d.c. between top line 2 and bottom line 4 (and also supplies steady current to the field of

---

* These are compensating fields in the amplidyne generator, connected in series with the amplidyne output current.

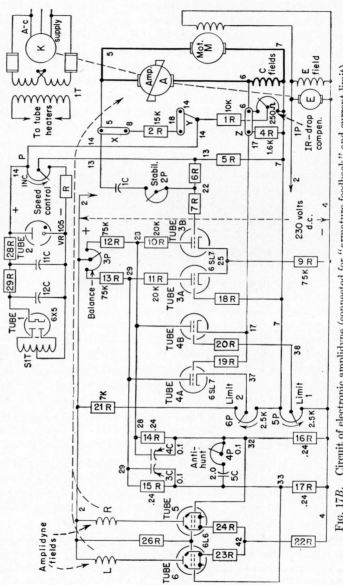

Fig. 17B. Circuit of electronic amplidyne (connected for "armature feedback" and current limit).

d-c motor M). The voltage divider $5P$–$6P$–$21R$ holds line 7 steadily about 50 volts more positive than 4.

17-2. Zero Amplidyne Voltage. Let us study Fig. 17B while speed control P asks for zero motor-M speed; slider 13 of P is touching at 14 so the "reference voltage" is zero. We shall see that 13 is at the potential of line 7. Since the tube-3A grid is always at 7 potential, the grids of both tubes 3A and 3B are now at this same potential; since their cathodes are together at point 25, these tubes have equal grid voltages. The current flowing through tube 3A, 11R and 13R is equal to the current through tube 3B, 10R and 12R, so points 28 and 29 are at the same potential; this voltage (28-to-4 or 29-to-4), applied across the duplicate voltage dividers 14R–16R and 15R–17R, places grid 32 at the same potential as grid 33, so the current through tube 5 and amplidyne field R is equal to the current through tube 6 and amplidyne field L.* These two field windings are alike, and are connected so that they oppose or buck; when they receive equal currents, their combined field strength is zero, so amplidyne A produces zero voltage, although A is being driven at full speed by a-c motor K.† Since the amplidyne produces zero voltage and current, its terminals 5 and 6 are at the potential of 7; therefore all points on the divider 2R–1R–1P are at 7 potential, so point 14 is holding 13 and the tube-3B grid 22 at 7 potential, as mentioned above.

Notice that tubes 3A and 3B act as a long-tailed pair (see Sec. 12-8). While their grids are at point-7 potential (50 volts above point 4), enough current flows through these tubes and 9R so that cathode 25 is raised about 51 volts above 4; each tube has a grid voltage of about -1 volt and passes about $\frac{1}{3}$ ma. Together they become a balanced d-c amplifier.

So long as no current flows in amplidyne A or its series fields, tubes 4A and 4B have no effect in Fig. 17B; line 17 is at 7 potential and connects to tube-4A grid and tube-4B cathode. Limit adjuster 5P is set at low potential to cut off tube-4B current; 6P raises the tube-4A cathode high enough to prevent any current flow.

---

*Tubes 5 and 6 are beam power amplifiers (described in receiving-tube manuals, also in Chap. 7 of Chute, "Electronics in Industry"); here each tube passes about 25 ma through its amplidyne field. Below these tubes, 23R and 24R are used for circuit stability, as is explained in Sec. 18-13.

† To make sure that the amplidyne fields L and R are exactly opposed, balance adjuster 3P is provided above tubes 3A and 3B; the 3P slider is adjusted until motor M does not turn.

282   ELECTRONIC MOTOR AND WELDER CONTROLS

17-3. Starting and Running of Motor M. If speed control P now is turned downward (top of Fig. 17B), a d-c voltage appears between 14 and P slider 13. Since 14 is yet at 7 potential, slider 13 is turned to a potential below 7; grid 22 of tube 3B also is made more negative, decreasing the flow of electrons through 9R, tube 3B, 10R, 12R and 3P. Since the voltage across 9R decreases, lowering cathode 25, this increases the flow of electrons through 9R, tube 3A, 11R, 13R and 3P. The voltage increases across 13R but decreases across 12R; point 28 rises and point 29 falls, both caused when P is turned, lowering the tube-3B grid.

As point 28 rises, the voltage divider 14R–16R raises grid 32 to increase the tube-5 current through the R field; similarly, when point 29 falls, the divider 15R–17R lowers grid 33 to decrease the tube-6 current through the L field.* Although the R-field current now may be only several milliamperes greater than the L-field current, this upsets the balance mentioned above, so that amplidyne A receives an exciting field strength and generates a voltage that makes its terminal 5 more positive than 6. This rise of point-5 potential also raises point 14; the system comes into balance when amplidyne A generates just enough voltage so that 14 is raised above 7 by an amount nearly equal to the 14-to-13 voltage selected by the turning of P.† So, whenever P is turned to a new speed position, the new amount of reference voltage 14-to-13 makes the tubes and the amplidyne respond until the feed-back voltage 14-to-7 (which is a direct indication of motor-M armature voltage) again nearly equals the 14-to-13 reference.

A small reversing switch may be used with the electronic amplidyne, to change the direction of turning of motor M. Not shown in Fig. 17B (but included in Fig. 17C), this switch is inserted at the right of tube 2; for reversed operation, this switch transfers point 14 to the negative side of tube 2, and connects R to the positive side, at 28R. Thus reversed, slider 13 becomes more positive as P is turned for higher speed; this turns on tube 3B, turning off tube 3A, so 28 falls and 29 rises. Tube 6 strengthens the L field while tube 5 weakens the R field. With L stronger than R,

---

* Capacitor 4C across 14R, or 3C across 15R, serves only to stabilize the circuit. Such action is described in Chap. 18, together with the 5C-4P antihunt circuit and the 1C-2P stability circuit.

† This electronic-amplidyne system produces amplification so that tube-3B grid 22 is lowered only $\frac{1}{10}$ volt in order to make amplidyne A apply 65 volts to motor M.

the difference now makes amplidyne A generate voltage that makes terminal 5 more negative than 6; this reversed armature voltage makes motor M turn in the reverse direction. The 5-to-7 armature voltage increases, driving point 14 negative until the 14-to-7 voltage again matches or offsets the 14-to-13 reference voltage of P.

17–4. IR-drop Compensation. The circuit action this far makes amplidyne A generate a constant voltage at any selected speed setting of P. If the load on the motor-M shaft increases, the speed of this d-c motor decreases because of the greater voltage loss in the motor's resistance (its internal IR drop). To offset this natural drop in speed, Fig. 17B makes use of the voltage across the C fields, produced by the motor current; part of this 6-to-7 voltage is selected by 1P, to hold constant the motor speed. When motor M is running forward, increased load makes the motor current drive point 6 and the 1P slider more negative than 7; this pulls point 14 more negative, so the tube circuits make amplidyne A generate greater voltage (more positive at 5) to bring 14 back to the point of balance. Thus the voltage applied to motor M is increased just enough to offset its natural slowing down due to load.

Similarly, when motor M is running in the reversed direction, increased load makes the motor current drive point 6 and the 1P slider more positive than 7 (although the A terminal 5 now is more negative than 7). As 14 is pulled more positive, amplidyne A is made to generate greater voltage, driving 5 still more negative to bring 14 back to the point of balance; the voltage applied to motor M is increased just enough to hold its speed constant despite load changes.

This compensation at 1P is not used when the electronic amplidyne is connected for tachometer feedback or constant-current control; in Figs. 17C and 17D, the 1P slider is set to touch at point 7.

17–5. Current Limit. To prevent motor M from drawing too much current while it is starting, stopping, or carrying heavy load, the entire voltage 6-to-7 across the C fields is used to control tubes 4A and 4B.

Let us think of motor M as turning only in the forward direction (with connections to P as are shown at the top of Fig. 17B). When the motor is rising to higher speed or has heavy shaft load, its armature current drives terminal 6 and point 17 more negative than 7. To limit this current, 5P (Limit 1) is set so that cathode

17 of tube $4B$ is driven down near to the potential of the $5P$ slider 38. As the motor current nears the desired limit, some current begins to flow through tube $4B$ and $12R$, so as to lower the point-28 potential; this decreases the tube-5 current through the R field, so that amplidyne A generates less voltage, decreasing the current that motor M can draw.

If P now is turned quickly to a lower-speed position (thereby decreasing the 5-to-6 voltage generated by A), M may act as a

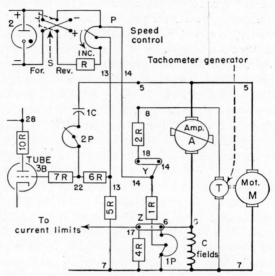

Fig. 17C. Electronic amplidyne connections for "tachometer feedback" and current limit.

generator to "pump back" current to A; this reversed current drives point 6 more positive than 7. Limit 2 ($6P$) is set to limit this current during slow-down; as the pump-back current nears the desired limit, point 17 and the tube-$4A$ grid are raised close to cathode 37. Current flows through tube $4A$ and $13R$, lowering the point-29 potential; this decreases the tube-6 current through the L field. It increases the difference between the R and L fields so that amplidyne A is made to generate greater voltage 5-to-6, thereby decreasing the current that motor M can pump back to A.

If now the reversing switch (mentioned in Sec. 17–3) reconnects P so that motor M runs in the opposite or reversed direction, the above actions are interchanged. During starting or carrying load,

THE ELECTRONIC AMPLIDYNE

the motor-M current drives point 6 more positive than 7; Limit 2 adjusts this maximum current. During slow-down (while motor M still turns in the reversed direction), 6 is driven more negative than 7; Limit 1 adjusts this maximum pump-back current.

17-6. Tachometer Feedback. To hold more constant motor speed during long running periods (such as may change the temperature, resistance and current of the motor fields) a tachometer generator may be added, to give the motor-speed signal to the electronic amplidyne. Driven by the motor-M shaft, this tachometer T is connected into the circuit as shown in Fig. 17C; jumper X is removed between 5 and 8. The voltage produced by this little d-c generator T increases as the motor speed rises; when motor M runs in the forward direction, the T voltage drives its terminal 8 more positive. If M is reversed, T also runs in the opposite direction and its voltage drives terminal 8 more negative.

The 1P slider is turned to its minimum, to touch at point 7, so that the voltage across the C fields has no effect on the motor-M speed. However, this 6-to-7 voltage is used (as in Sec. 17–5) to limit the motor-M current to the amount set by Limit 1 or Limit 2.

With the reversing switch S thrown to the left in Fig. 17C, for forward operation, point 13 becomes more negative as P is turned downward to raise the motor-M speed. As in Sec. 17–3, the 14-to-13 voltage set by P becomes the reference voltage. By forcing 13 and the tube-3B grid 22 more negative, current is increased in tube 5, decreased in tube 6, so that amplidyne A generates voltage that is positive at terminal 5, and motor M runs forward, as was described previously. However, the motor-armature voltage 5-to-7 has no effect on the circuit operation. Instead, the motor speed rises until its tachometer T generates enough voltage to raise terminal 8 and thereby to make the 14-to-7 voltage nearly equal to the reference voltage 14-to-13. Instead of motor-armature feedback (as in Fig. 17B), the tachometer voltage is the feed-back signal to match the selected speed setting of P.*

17-7. Constant-current Regulation. The electronic amplidyne described in Sec. 17–1 may be used to control motor M so that it draws a constant amount of armature current (as set by adjuster P); motor M therefore delivers constant torque to the driven load,

---

* Since the tachometer generator has permanent-magnet fields and its generated voltage is not affected by motor-field temperature, the T voltage is a better long-time indication of speed of the motor that drives it.

while the motor speed may vary. Connections for such constant-current operation are shown in Fig. 17D. Notice that jumpers Y and Z now connect 18 to 17 and connect 14 to 6; jumper X connects 5 to 8 (as in Fig. 17B). Instead of controlling speed, adjuster P now selects the amount of motor-M current and the torque at the motor shaft. The turning of P controls tube $3B$ as in Sec. 17-3, to make tubes 5 and 6 strengthen one field and weaken the other until amplidyne A produces voltage to force the required current through motor M. However, because of the changed position of jumper Z, the feed-back signal does not come from the armature voltage at point 5 (as in Fig. 17B) or from a speed signal at point 8 (as in Fig. 17C); instead, the feed-back signal (between points 14 and 7 in Fig. 17D) comes directly from the voltage drop across the C fields. This C voltage increases as the motor-M current increases and is a direct signal that indicates the amount of M armature current. Meanwhile, the motor-armature voltage 5-to-7 is connected through jumpers X and Y so as to operate the limit tubes $4A$ and $4B$ to prevent too high motor speed.

In previous sections and in Figs. 17B and 17C, the turning of slider P to lower potential (to increase motor speed in the forward direction) makes amplidyne A generate voltage so that terminal 5 becomes more positive; positive potential is fed back at point 14. However, when jumpers Y and Z are connected as in Fig. 17D, the feed-back potential comes from point 6; but 6 becomes more *negative* if motor current flows when forced by positive terminal 5. So, for this constant-current regulation, the amplidyne fields also are reconnected so that tube 5 now supplies the L field, and tube 6 supplies the R field. The "Forward" position of reversing switch S refers to forward motor current and not to forward rotation.

With switch S in the "Forward" position in Fig. 17D, the tube-$3B$ grid 22 is driven more negative as P is turned to increase the motor-M current. As is described in Sec. 17-3, the tube-$3B$ current decreases, raising the potential of point 28, thereby raising grid 32; but tube 5 now strengthens the L amplidyne field. At the same time, the tube-$3A$ current increases, lowering points 29 and 33 so that tube 6 weakens the R field. This difference between R and L makes amplidyne A generate voltage that is negative at terminal 5; the motor-M current flows in the direction that makes the C-field terminal 6 more positive than 7. This current becomes steady when it produces enough 6-to-7 voltage to match the reference voltage 14-to-13 set by P.

THE ELECTRONIC AMPLIDYNE   287

Let us see how the motor-M speed changes if its shaft load changes. Suppose that this motor is rated to carry a full-load current of 5 amperes at 230 volts and full speed of 1750 rpm.* If P now is set to hold a constant current of 3 amperes, the motor-M speed rises until its shaft load makes the motor draw 3 amperes.

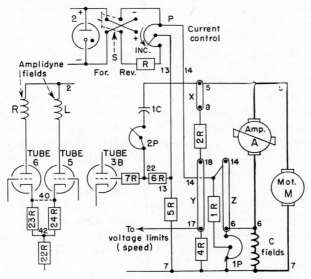

Fig. 17D. Electronic amplidyne connections for constant-current regulation and speed limit.

If the required load is too light, the motor rises to its top speed; if the load still does not make the motor draw 3 amperes, the circuit cannot regulate and the motor draws less current. When the shaft load increases so that the motor tries to draw more than 3 amperes, the motor slows down until the required shaft load makes the motor draw just 3 amperes. If the shaft load still is too great, the motor may stall; yet it draws only 3 amperes, or less than its rated full-load current.

17-8. Speed Limiter. If motor M in Fig. 17D usually runs between 600 and 1000 rpm when loaded to 3 amperes, one may not wish to let the motor speed rise to 1750 rpm each time its shaft load is removed. Limit 1 (5P in Fig. 17B) may be set to limit this motor speed. At that amount of armature voltage

* The motor base speed is 1750 rpm, since exciter E supplies full-field current at all times to motor M. See Sec. 13-2.

(such as 160 volts, 5-to-7) that holds 3 amp in the motor at 1050 rpm, $5P$ is set so that tube $4A$ just fails to pass current. Then, if the motor load lessens (so that the C voltage 6-to-7 decreases and amplidyne A starts to increase its voltage to raise the motor speed), the increasing 5-to-7 voltage drives point 17 more negative, lowering the tube-$4B$ cathode. This lets current flow through tube $4B$ and $12R$, preventing a further rise at points 28 and 32, so that tube 5 cannot further increase the strength of the L field. The amplidyne voltage remains close to 160 volts, so the motor-M speed remains below 1100 rpm, because of the voltage-limit action of tube $4A$.

If switch S is thrown to the "Reverse" side in Fig. 17D, P increases the motor current by moving point 13 to higher potential. Amplidyne A now generates voltage that is more positive at terminal 5; motor M now turns in the opposite direction and the motor current makes point 6 more negative than 7, driving point 14 negative also until the 14-to-7 feed-back voltage matches the 14-to-13 reference voltage. To limit the top speed when load is removed from the motor, Limit 2 is set so that tube $4A$ passes current when the increasing 5-to-7 voltage raises point 17 and the tube-$4A$ grid; tube 6 is prevented from strengthening the R field, so the amplidyne cannot increase its voltage further, to raise the motor speed.

17–9. Stability and Antihunt Circuits. The operation of the Fig. 17B circuit is explained above without including the effect of resistors $23R$ and $24R$, capacitors $1C$, $3C$, $4C$, $5C$ or adjusters $2P$ and $4P$, except to mention that they stabilize the circuit or prevent the system from hunting.

Most descriptions of such added parts will merely say that $1C$ and $2P$ together produce a "phase-lead" action; that $23R$ is used to decrease the time constant of amplidyne field L; that $5C$ and $4P$ produce a "phase-lag" action; or that $5C$ and $4P$, combined with $3C$, $15R$, $4C$ and $14R$, comprise a "conventional notch network." So that these statements may have more meaning, and so as to indicate the reasons why such parts are added, the next chapter is presented. For example, we shall see that the "notch network" gets its name because, when correctly designed, it causes the shape of a notch to appear in the Bode diagram of the system, such as is shown in Fig. 18I.

CHAPTER 18

STABILITY OF A CLOSED-CYCLE SYSTEM

Each motor control in this book is a closed-cycle system. As is described in Sec. 12–5, a dial may be turned to select a new position or output voltage or speed; when the motor has caused the load to reach this new position, voltage or speed, an electrical signal is sent back to report "mission accomplished." This closes the cycle, and the system becomes self-regulating.

18-1. The Problem of Hunting. To obtain close accuracy in such regulating systems, the input signal (set by the dial) is greatly strengthened by the use of tube amplifiers, amplidynes, etc.; only a small part of the system's large output is fed back to balance against the low-level input signal. Most motor-control systems include such amplification. However, this often makes the system waver or hunt, so that the motor refuses to "settle down" to a fixed speed, or to hold steadily the desired position or tension. Usually such hunting can be reduced by decreasing the amount of amplification; since this also decreases the system's accuracy, it is not the best answer to the problem. Instead, the system designers have learned how to insert or add antihunt devices such as special transformers or certain combinations of capacitors and resistors; the only purpose of these added parts is to stabilize the system— to prevent it from hunting.

When an electronic amplifier is used as part of the closed-cycle system, the antihunt devices are placed in this electronic portion, for here the needed parts (capacitors, etc.) may be much smaller and are more easily changed to attain the desired results. Such antihunt circuits are mentioned in earlier chapters. To reduce hunting or "overshoot" in Sec. 12–9, we increase the dead zone of the positioning control; this merely reduces the system's sensitivity and accuracy. Then, in the loop control of Sec. 13–7 and Fig. 13G, capacitors $3C$ or $4C$ are added as a "slow-down" device, and $14C$–$18R$ together for "speed-up." Similarly, Sec. 14–4 and Fig. 14D include $3C$–$14R$ for slow-down, $5C$–$6R$ for speed-up; Sec. 15–19 and Fig. 15J include $2C$–$10R$ for slow-down, with

290 ELECTRONIC MOTOR AND WELDER CONTROLS

$14C$–$54R$, $10C$–$49R$ and $12C$–$53R$ for speed-up. Such names are in use because the added slow-down circuit makes the hunting or oscillation become less frequent until it disappears; the added speed-up circuit makes the system respond at a faster rate so that the hunting disappears.

18–2. The Intent of This Chapter. As the result of many trial circuit changes, and careful analysis of the results, designers have learned how to "antihunt" a system so that most of the corrective devices may be included in the original design. Electrical engineers now have ways to calculate in advance whether or not a system will hunt. Such methods in detail are beyond the scope of this book.*

System designers, highly trained engineers, may understand the theory of these antihunt methods. Many of the circuit constants needed in such analyses are obtained only by careful tests and study. We shall not try to explain such system-stability studies. Instead, a quick survey or sketch of one of the engineers' ways of describing stability may be helpful to users of motor-control systems; by being familiar with the Bode diagram, such as an engineer may draw so as to study the system, the field engineer may see better where certain antihunt devices fit into the control scheme. Bode diagrams appear in Figs. $18G$ and $18J$, to be discussed later; such a diagram is based on many short cuts and assumptions, and applies only to certain simple closed-cycle systems.

18–3. A Cause of Hunting. As a probable cause of hunting, motor-control systems often contain two or more major phase lags, the result of inductive and electromechanical time constants. To explain this statement, Fig. $18A$ shows a system that includes an electronic amplifier and a rotating amplidyne, similar to the electronic amplidyne of Chap. 17; the amplidyne supplies current to the field of a larger generator. A feed-back signal from the generator's output voltage returns to the amplifier, closing the cycle. Each field winding in this system is an inductive load, for the field current lags behind the applied voltage whenever that voltage is changing. Whenever the system is steady, the current in the amplidyne control fields closely matches the voltage applied to these fields by the amplifier tubes; however, if the tubes turn

---

* The theoretical background of the Nyquist approach and the Bode attenuation diagram is summarized in Sec. 4 of W. D. Cockrell, "Industrial Electronic Control," 2d ed., McGraw-Hill Book Company, Inc., New York, 1950.

STABILITY OF A CLOSED-CYCLE SYSTEM

on and off and rapidly change the control-field voltage, the field current lags behind these voltage changes. This phase lag may be as much as 90 deg. This lagging current next causes voltage to be produced at the amplidyne's quadrature circuit; this circuit also is inductive, so there is another phase lag (of current behind the voltage), perhaps as much as 90 deg. This lagging quadrature current now makes the amplidyne generate voltage which is applied

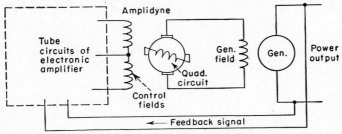

FIG. 18A. A closed-cycle system showing probable sources of hunting.

to the large generator field; this large field winding is highly inductive, so its current also lags. With three of these phase lags combined in the one system of Fig. 18A, it is seen that the feed-back signal may lag a total of 180 deg or more behind the voltage produced by the amplifier; if the amplifier provides a circuit gain greater than 1.0, then such a signal, 180 deg out of phase, easily makes the circuit oscillate so that the whole system hunts.* These phase lags receive further attention below.

18–4. Signal Gain at Various Frequencies.

When an engineer studies a closed-cycle system, to learn how well that system responds to various signals, he may draw a curve whose shape tells how much the signal gains or loses in strength for each frequency or kind of input signal that may be applied to the system. In simple form, such a curve may appear like Fig. 18B. At A, while the signal changes slowly (at less than 5 pulses per second), Fig. 18B shows that the system strengthens the signal so that the gain is

---

* Any system will have 180 deg shift at some signal frequency, but usually the gain is so low that the system does not hunt. If a certain frequency is reinforced or built up within the system so that it feeds back with enough strength (greater than 0 db) to cause oscillation, an antihunt circuit may be added to weaken the signal at that frequency. This will be a series antihunt circuit, inserted into the system so that the entire signal must pass through it, as contrasted with a feed-back antihunt circuit that by-passes part of the main circuit.

about 5 times; for each volt input to the system, it delivers 5 volts output. However, if the signal changes more rapidly, say 30 times per second, the curve shows (at B) that the system's gain has decreased to 4; or, changing 50 times per second, the gain is only 2 (at C). As the signal frequency increases, this system's output decreases, mainly because the various field currents lag farther behind their voltages.

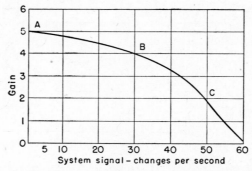

FIG. 18B. A simple curve may show how a signal (at various frequencies) gains strength in the system.

Instead of making his curve on a simple chart like Fig. 18B, the engineer may use a logarithmic scale as in Fig. 18C, so that wider ranges of gain and frequency can be shown. Also, he may use straight lines when drawing such curves, instead of a longer point-by-point method. Although the lines are marked differently, the idea is the same as in Fig. 18B. If needed, the footnote shows the decibel (db) values for various gains, and also shows that 10 radians per sec is about equal to 60 cycles or pulses per sec.*

Gain	Db	Gain	Db
0.5	−6	32	30
1	0	56	35
2	6	100	40
4	12	177	45
7	17	316	50
10	20	560	55
18	25	1000	60

The decibel (db) = 20 $\log_{10}$ gain

Frequency in radians per second, as in Fig. 18C, is equal to cycles per second multiplied by 2π, or 6.28. So radians = $2\pi f$ (cycles), and

10 radians per sec = 62.8 cycles per sec

STABILITY OF A CLOSED-CYCLE SYSTEM 293

In Fig. 18C, the heavy lines show how well the amplidyne and generator system (of Fig. 18A) can perform; the upper line shows the system with its usual high gain (such as 316 times, or 50 db) while the lower line represents a similar system that has very low gain (such as 10 times, or 20 db). We shall see that the break or bend at J or M is caused by the large generator field, while the amplidyne's quadrature circuit causes the bend at K or N. To tell where to draw these bends, the engineer calculates the time constant (T.C.) of each field winding, as is outlined below. The

	Amplidyne control field	Amplidyne quadrature circuit	Generator field
Resistance R, ohms..	1600	2	50
Tube ohms*........	1700		
Inductance L, henrys	55	0.2	25
Time constant, L/R .	0.0167	0.1	0.5
1/T.C.............	60	10	2

values of 1/T.C. directly locate the bends in Fig. 18C; so, for the large generator of Fig.18 A, 1/T.C. = 2.0, and J is located at 2 radians and 50 db. The system's greatest gain, 50 db, is calculated from the amplifier-tube gain and the amplidyne's voltage gain; such calculations are not included here.

To the right of J or M, the heavy line slopes downward at the rate of 20 db per decade,† indicating that the system now loses its gain at the same rate as the signal frequency rises. This slope (as between J and K) is also called a unit slope; this unit slope (or slope of 1) will appear often in later discussion. On this sloping line, K is located at the vertical line 10, since the preceding table shows that the next larger value of 1/T.C. is 10, for the quadrature circuit.

Beyond the second break (or to the right of K) the line has a 2-unit slope; line K-to-S is drawn so that S is 40 db below K, and at a frequency of 100 (or 10 times the frequency at K). On this

---

* Tube resistance is added into the field that the tube controls.

† A decade spans a frequency range of 10 to 1. To draw line JR in Fig. 18C, locate R on a line 20 db below J; since the frequency at J is 2, locate R at 2 × 10 or a frequency of 20. This slope is also 6 db per octave (falling 6 db for each doubling of frequency, as from 2 to 4, or from 5 to 10).

2-unit-slope line, the bend at L is located at the vertical line 60, to match the 1/T.C. value shown for the control field in the table.

18-5. Study of the Bode Diagram.* In Fig. 18C, the straight sloping lines follow closely the true response of the system (which is shown by the dotted curve, and can be proven by test). Moreover, this curve indicates the amount of phase lag in the system, for tests will show that, at bend J or M, the system output lags 45 deg behind the input signal; at bend K or N, the phase lag is 135 deg, increasing to 225 deg at L. When we recall (from Sec. 18-3) that a system will hunt if the feed-back signal lags by as much as 180 deg (when the gain is more than 0 db), we see

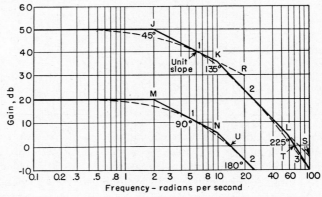

Fig. 18C. A Bode diagram showing gain vs. frequency, on logarithmic scale.

that the high-gain, or upper, line in Fig. 18C indicates an unstable system, for the phase lag has reached 180 deg before the sloping line crosses the 0-db line at T. However, the lower line shows how this system may become stable when the system gain is reduced, for now the phase lag is greater than 135 deg but is not yet 180 deg at the point where the line reaches 0 db, at U. After drawing the system action as in Fig. 18C, the designing engineer usually may expect that the system will hunt, have poor response or be unstable if the sloping line has more than a unit slope when it crosses the 0-db line. Therefore, to stabilize the high-gain system of Fig. 18C, he will add antihunt circuits that will change the shape of the system curve until it has a unit slope at 0 db. Such methods and results will be shown in Figs. 18G and 18J.

While adding antihunt circuits, the engineer tries to keep high

* Based on the work of Dr. H. W. Bode (Bō'dey) of the Bell Laboratories.

system gain, especially at the lower frequencies, for this gain (at the left in Fig. 18C) provides the high sensitivity and steady-state accuracy desired in the average industrial closed-cycle system. Most often the antihunt circuits are merely combinations of resistors (R) and capacitors (C), known as RC circuits; several of these are next described.

18–6. The Phase-lag Network.* When a resistor and a capacitor are connected into a system as shown in Fig. 18D, the effect is like that of a field winding (as mentioned in Sec. 18–3); a phase lag is produced, thereby decreasing the system's output as the signal frequency is increased. If the input voltage has very low frequency (or is d.c.), we know that capacitor 1C may have little effect in Fig. 18D; so little current flows in 1C that the output voltage may equal the input voltage. (Since output/input = 1,

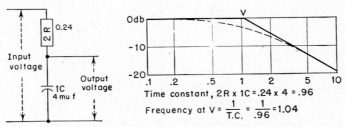

Fig. 18D. A phase-lag network and its diagram.

we say that this network has a gain of 1; this corresponds to 0 db.) However, as the system frequency increases, more a.c. flows in 1C and 2R, causing a voltage drop across 2R; the output voltage decreases (along a unit slope or a slope of 1), causing a loss in system response. At very high frequency, 1C becomes almost a short circuit so that nearly all the input is lost across 2R; the system curve continues its downward slope to large (−db) values. The bend in the curve is centered at that vertical frequency whose value is calculated by 1/T.C. or 1/(2R × 1C). For the 0.24-meg and 4-mu f values shown, this bend at V is at 1.04 radians per sec.

If this network is added into the system near a tube (as in Fig. 18E), with the capacitor at position XC, the conditions are the same as in Fig. 18D. However, if the capacitor is placed instead between tube grid and anode, its 4 mu f may be more effective at 1C than at XC. If the tube has a gain of 8, 1C acts like 9 ×

---

* This is also called the "integrating network" or "slow-down circuit." "Network" refers to any group of electrical parts connected together.

4 mu f,* and the bend is moved to the left accordingly (to 1.04/9, or 0.12 rad). Of course, if the tube has a gain less than 1 (as may apply to a cathode follower), the capacitor may have the same effect whether it is connected at $1C$ or at XC.

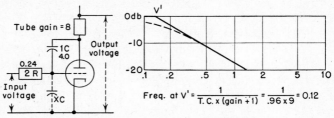

Fig. 18E. Phase-lag network applied to a tube circuit.

If now a resistor $1R$ is added in series with $1C$, as is shown in Fig. 18F, this network causes *limited* lag or slow-down. The first bend, at V, caused by $2R$ and $1C$, is the same as in Fig. 18D. At high frequencies where $1C$ becomes nearly a short circuit, $2R$ and $1R$ now act as a voltage divider so that the output voltage becomes a fixed portion of the input voltage; there is no further loss in output at still higher frequencies so the system curve again becomes horizontal. The location of the second bend, at W in Fig. 18F,

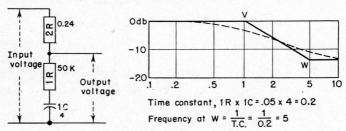

Fig. 18F. Limited phase-lag network.

depends on the values of $1R$ and $1C$, and occurs at a frequency of 1/T.C. or $1/(1R \times 1C)$. For 0.05 meg and 4 mu f, the bend is at 5 radians per sec.†

In any phase-lag network, notice that the capacitor is in a circuit *across* the output voltage; since the capacitor's voltage does

---

* The multiplication factor = gain + 1.

† By calculation, it may be seen that the horizontal line (to the right of W in Fig. 18F) is lowered by 13.7 db. The voltage-divider action of $1R$ and $2R$ makes the output voltage = 0.05/0.24, or 0.21, times the input voltage. A gain of 0.21 corresponds to $-$ 13.7 db.

not change instantly, it "braces" the circuit against sudden changes, slowing the circuit action.

18–7. Stabilizing with a Lag Network. Let us see how the lag network of Fig. 18F is used to stabilize a closed-cycle system. In Fig. 18G, the solid line is drawn for the system studied previously in Figs. 18A and 18C; however, the gain at the left in Fig. 18G is adjusted to 25 db, so that the whole solid line M–N is raised 5 db higher than before. Since the second bend, at N, now occurs far above the 0-db line, this system's phase may lag by nearly 180 deg at 0 db, so hunting may be expected.

For comparison, the bending lines of Fig. 18F are drawn again at V–W in Fig. 18G. If 1R, 1C and 2R now are connected into the system (probably within the electronic-amplifier circuit sketched

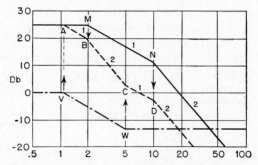

Fig. 18G. Network of Fig. 18F used to stabilize a system.

in Fig. 18A), the effect of this added network can be shown in Fig. 18G. Beginning at the left at a system gain of 25 db, the bend at V causes a similar bend at A; the system gain decreases, following the downward slope of 1 toward B. The bend at B is caused by the large generator field at M; the system quickly loses gain, following a slope of 2. At C, the effect of 1R is noticed; the end of the downward slope at W causes a similar decrease of slope at C. Notice that the line C–D has a slope of 1 and crosses the 0-db line; this fulfills the main requirement in the Bode diagram, to show that this corrected system now may be stable.* The lag at N, caused by an amplidyne field, makes a similar bend at D so that the system line returns to a slope of 2, but this may be too late to affect the system stability.

* Stability depends also upon phase shift, calculated by means not included here.

18–8. The Phase-lead Network.* When two resistors and a capacitor are connected into a system as shown in Fig. 18H, a phase lead or a speed-up action is produced, thereby increasing the system's output at higher signal frequencies.† At a low signal frequency, capacitor 2C passes very little current, so 1R and 2R act as an ordinary voltage divider; the output voltage is a fraction of the input voltage. If, as in Fig. 18H, the output is 0.05/0.24, or 0.21, times the input, the signal gain of this network at low frequency is -13.7 db (so that this is actually a loss). However, as the signal frequency rises, more a.c. passes through 2C, so that the voltage across 2R decreases, thereby increasing the 1R and output voltage; the network lets the gain increase. Finally, at

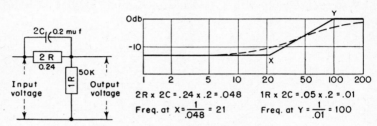

Fig. 18H. Limited phase-lead network.

high frequencies that make 2C act as a short circuit across 2R, the output voltage equals the input voltage, so the signal gain becomes 1, or 0 db. The bend at X is located by $2R \times 2C$; at Y by $1R \times 2C$, as is calculated in Fig. 18H.

18–9. The Notch, or Lag-lead, Network. For use in stabilizing a high-gain circuit, the phase-lead network above usually is combined with a phase-lag network; this combination is called a "notch network." The lead network of Fig. 18H is made into a notch network merely by inserting capacitor 1C in series with 1R, as is shown in Fig. 18I. Since 1C is 4 mu f as in Fig. 18F, the slanting-line diagram of this network is seen to be like that of

---

* Also called the "differentiating network" or "speed-up circuit."

† Notice here that the capacitor is in a circuit *in series* with the signal; the signal must pass through 2C or through resistor 2R that parallels it. For an instant after the signal is applied at the input, the 2C voltage remains unchanged, making the entire signal appear at the output; then, as 2C changes its charge to match the signal, only part of the input signal remains at the output. In this way 2C causes a forcing action each time that the input signal changes.

Fig. 18F added to that of Fig. 18H.* This combined diagram shows (at V–W–X–Y) the notch shape from which this network is named.

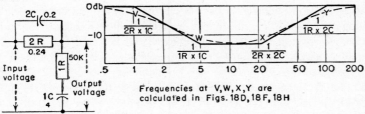

FIG. 18I. A notch or lag-lead network.

18–10. Stabilizing with a Notch Network. To show how this network is applied, Fig. 18J shows again the solid line J–K–L from Fig. 18C, showing the performance of the system of Fig. 18A.

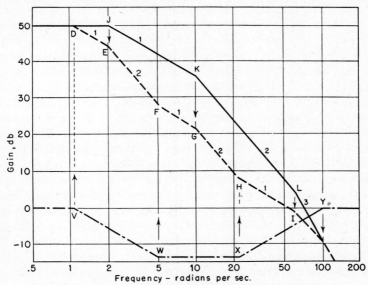

FIG. 18J. Notch network of Fig. 18I used to stabilize a system.

For comparison, the bending lines V–W–X–Y are added from Fig. 18I.

If the notch network (1R, 1C, 2R, 2C of Fig. 18I) now is connected into the system's electronic-amplifier circuit, the effect of

* This addition of separate lag and lead diagrams to form a notch is approximate only.

this added network can be shown by the dashed line D–E–F–G–H–I in Fig. 18J. The first bend, at D, corresponds to bend V of the notch network. The effect of the large generator field (at J) causes a second bend at E so that the system gain decreases rapidly along a slope of 2. The notch-network bend W causes the bend at F, decreasing the slope. However, the K bend (of the amplidyne's quadrature circuit) causes another downward bend at G. A short distance above the 0-db line, another bend H is caused by the notch network (at X); this decreases the slope so that the line H–I has merely a slope of 1 when it crosses the 0-db line. This fulfills the main requirement in this Bode diagram, to show that this corrected system now may be stable. The bend at L, caused by the amplidyne's control field, makes the similar bend at I; the system line returns to a slope of 2, and later to a slope of 3 (because of the notch network, at Y). If these steep slopes do not occur near the 0-db line, they do not indicate instability.

18–11. A Notch-network Variation. The circuits discussed in Chaps. 17 and 19 include notch networks that have the form shown in Fig. 18K.* These networks are used with pairs of tubes through which the current passes to the control fields of an amplidyne, or to the d-c windings of saturable reactors. Such tubes are connected so that they have a gain of about 10. Since 1R and 1C in Fig. 18K are connected between the grids of the two tubes, they may not appear to be part of a notch network. However, the circuits of Fig. 18K act as if they were regrouped to control each tube separately, as shown in Fig. 18L. Note that the single 2-mu f capacitor (1C in Fig. 18K) is replaced by two 4-mu f capacitors in series, to have the same electrical value; similarly the single 0.1-meg resistor 1R is split into two 50-kilohm resistors. When the tube-1 grid rises in a circuit such as Fig. 18K, the tube-2 grid falls an equal amount, so the center-point potential between the two grids (shown grounded through capacitor C in Fig. 18L) remains unchanged. Thus tube 1 has its notch network equal to 1R, 1C, 2R, 2C in Fig. 18L.

*The parts shown in Fig. 18K compare with those in similar circuits, as follows:

Fig. 18K	Fig. 17B	Fig. 19D
1R	4P	17R
1C	5C	8C
2R	15R	10R
2C	3C	11C
Tube 1	Tube 6	Tube AA

STABILITY OF A CLOSED-CYCLE SYSTEM

Since the resistor and capacitor values in Fig. 18L are the same as are used in Sec. 18–9 and Fig. 18I, the notch network of Fig. 18K has the same effect as that analyzed in Fig. 18J.

18–12. Stability Comments. Only a few samples of system instability are given above, with approximate methods for improving the stability. A more complete study is not intended here.

Each system's tendency to hunt will depend on the amount of amplification or gain, and upon the time constants of fields or other phase-lag elements. Usually a system is more difficult to stabilize if its inductive elements (such as field windings) have time constants that are close together in value. Time constants of some circuit elements are difficult to determine and are variable; such data may not be readily available.

The location and shape of the notch in the Bode diagram may be changed by adjustment of the stability, sensitivity or antihunt dials in the circuit.

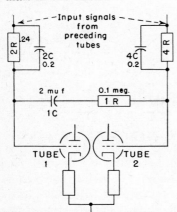

FIG. 18K. A form of notch network applied to pair of tubes.

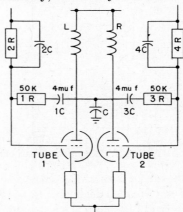

FIG. 18L. Modified form of network shown in Fig. 18K to show notch-network constants.

When the closed-cycle system includes a motor or similar device that is used to keep some object in a desired position, the Bode diagram of that system starts (at low frequency) with a downward slope of 1 (since the position at any instant represents an integration or totaling of all the preceding motor actions).

18–13. Cathode Resistors for Stability. In many electronic circuits, a resistor is inserted in the cathode circuit of a tube to improve the circuit stability. Several examples may explain the different actions thus produced.

Part (a) of Fig. 18M shows a cathode follower (from Fig. 13G). So connected, the tube has a gain of less than 1; aside from this, it does not help stability. In (b) a more common tube connection supplies output at the tube anode. The addition of cathode resistor 1R decreases the tube gain to increase stability (for, as the grid potential rises, to increase the anode current, the cathode potential rises also).

Part (c) shows the tube-$F2$ circuit from Figs. 16F or 16H. Here a voltage is inserted across 12R. As the tube-$F2$ grid potential rises, increasing the current through tube $F2$ and the d-c winding of reactor SX, greater current is supplied to the load (which is a motor-field winding) and passes also through 12R. This increased load current increases the voltage across 12R, raising the cathode

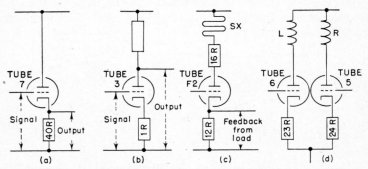

Fig. 18M. Examples of cathode resistors affecting system stability.

of tube $F2$ so as to decrease the grid-to-cathode voltage. This action (called degenerative feedback) helps to stabilize the circuit.

Part (d) shows the left-hand part of Fig. 17B. Resistor 23R is added in series with tube 6 and the amplidyne field so as to increase the resistance of this circuit and thereby to decrease the time constant of the amplidyne field; such a lowered time constant generally makes the system more easily stabilized. Connected in the cathode circuit, 23R may have greater effect than if it were connected in the tube-6 anode circuit; if tube 6 has a gain of 2, a 100-ohm resistance at 23R gives the stabilizing effect of a 300-ohm resistor added in the anode circuit.

As is shown also in Fig. 17D, these resistors 23R and 24R may be bridged by a jumper at 40 connecting the tube cathodes; this decreases the effect of these resistors, holding them in reserve unless needed for stability.

CHAPTER 19

THE PHOTOELECTRIC TRACER

To control the forward or sidewise movement of gas cutting torches so that they may cut irregular shapes from steel plate, such torches may be connected mechanically to a tracing head, as is shown in Fig. 19A. Within the tracing head, several phototubes "watch" the desired pattern on the table beneath, and make

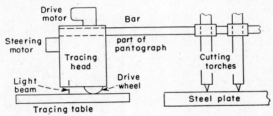

FIG. 19A. Cutting torches moved by tracing head that follows a template.

a steering motor turn the whole tracing head so as to follow along a black-line pattern or template; the equipment acts as an "electronic bloodhound."

19-1. Electronic Tracer (CR7512-B100). The electronic control cabinet is shown in Fig. 19B; Fig. 19C shows the tracing head, mounting several small motors. The circuit diagram appears in Fig. 19D. Beneath the tracing head is a drive wheel that rests upon a transparent sheet protecting the template surface; a driving motor turns this wheel at constant speed (as set by a governor on the motor shaft) so that the whole tracing head moves horizontally at the desired cutting speed. Through a mechanical pantograph, the cutting torches are made to follow exactly every movement of the tracing head. A small but strong light beam shines from the tracing head down onto the pattern or template, which may be a white paper cutout resting on a black background, or may be a black line $\frac{1}{8}$ in. wide, drawn on a white background to match the shape of the pieces to be cut. The light reflects up from this pattern to reach several phototubes, whose output is amplified so as to control a steering motor; by turning forward or backward, this motor changes the direction followed by the turning drive wheel. This wheel and the tracing head are steered in such direction that the light beam follows along on the edge of the template; half the light shines on the black line, half on the white background.

Briefly, if the head moves slightly off the black line, so that more of the light beam shines onto the white background, more light is reflected to the phototubes. This increases the current in tubes C and AA but decreases the current in tubes CC and A. From

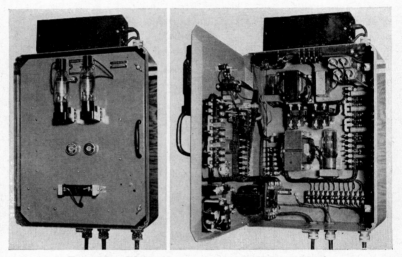

FIG. 19B. Electronic-tracer control (CR7512–B100).

the a-c supply line, current now flows in one direction through tube 2 and the steering-motor armature; the motor changes the direction of the drive wheel so as to steer the tracing head back onto the black line.

In the upper half of Fig. 19D, all circuits operate on alternating current. From this a-c supply, current passes through a constant-voltage transformer and $106T$, is rectified by tube B and is filtered by reactor $3X$ and capacitors, so that a constant 340 volts d.c. is supplied to the lower part of Fig. 19D, between lines 17 and 9. Tube B also supplies constant field current for the steering motor M. When a-c power is first applied, the tubes are warmed, through $102T$; soon the time-delay relay picks up relay CR, whose

THE PHOTOELECTRIC TRACER

FIG. 19C. Motor-driven tracing head controlled by light beam.

306 ELECTRONIC MOTOR AND WELDER CONTROLS

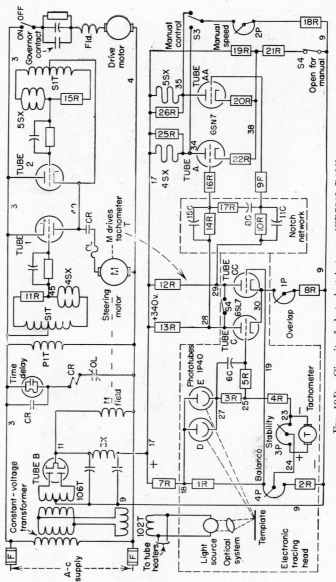

Fig. 19D. Circuit of photoelectric tracer (CR7512-B100).

THE PHOTOELECTRIC TRACER

contact connects the motor-M armature to tubes 1 and 2. These thyratrons are connected back to back; neither tube passes current so long as the scanning head is in its correct position above the edge of the template. If tube 1 alone fires, electrons flow from top line 3 through tube 1 and the motor-M armature to line 4. Since this current consists of half-wave pulses, all in the same direction, the d-c motor M turns in one direction. (This action is described in Sec. 14-6 and Fig. 14E.) However, if tube 2 fires, current flows through the armature during the half cycles of opposite polarity so that motor M turns in the opposite direction.

Meanwhile the drive motor (at the right in Fig. 19D) may run steadily as an a-c series motor; its speed-governor contact closes to apply full voltage to the motor, or, if the speed rises, the contact opens to decrease the motor voltage and slow the motor. The drive wheel turns steadily; the electronic circuits "say" in which direction this turning wheel shall travel.

19-2. On-line Operation. As in driving a car, there is no need to move the steering wheel so long as the car or the scanning head is moving in exactly the desired direction. If no correction is needed, the steering motor M does not turn, so tachometer T (at lower left in Fig. 19D) does not turn or produce voltage. Phototubes D and E receive enough light so that point 25 (on the voltage divider consisting of the phototubes, 3R, 4R and 3P) is at the same potential as the fixed point 19 (on the divider 7R-1R-4P-2R). Since 25 and 19 set the grid potentials of tubes C and CC, whose cathodes are together at 30, these tubes pass equal currents through the equal resistors 12R and 13R; therefore their anodes 28 and 29 are at equal potentials, so the grids of tubes A and AA also are at equal potentials.

Below tube CC, adjuster 1P may be turned clockwise to decrease the resistance in the cathode circuit, thereby lowering the cathode-30 potential; this increases the current through tubes C and CC, lowering equally the potentials of anodes 28 and 29. In this way the grids of tubes A and AA are lowered together until each tube passes very little current (for the tube cathodes are "tied" to a fixed potential at point 38 on the divider 19R-21R); these currents in 4SX and 5SX are so small that these reactors have large voltages across their a-c windings, thereby permitting thyratrons 1 and 2 to pass very little current.* Overlap adjuster 1P

---

* This phase-shifting action by 4SX is explained in Sec. 13-6. In the tube-1 grid circuit, transformer winding $S1T$ applies about 60 volts a.c. across

is set so that tubes 1 and 2 have very little blue glow. Balance adjuster $4P$ is turned until the glow in tube 1 is equal to the tube-2 glow, so that motor M does not turn.

19–3. Line-follower Action. When the black line or edge of the pattern turns or changes direction, the scanning head continues to move in its original direction until its light beam travels farther onto the black line or farther off the black line. Suppose that it moves off the line so that the white background reflects more light to the phototubes. At once phototubes D and E pass greater current, in Fig. 19D; more electrons flow from the $4P$ slider through $3P$, $4R$, $3R$ and the phototubes, increasing the voltage across these resistors. Since this raises the potential of point 25, more current may flow through tube C and 13R. Tubes C and CC act as a long-tailed pair, so less current flows through tube CC and 12R. We see that increased light on the phototubes lowers anode 28 and raises anode 29; current decreases in tube A and 4SX but increases in tube AA and 5SX. Thyratron 1 is turned off completely, but more current flows through thyratron 2 and the motor-M armature; motor M turns the vertical shaft that mounts the drive wheel—it steers the drive wheel in a new direction that aims to bring the scanning head's light beam back onto the black line.

Similarly, if the black line turns so that the light beam moves farther onto the black line, the phototubes receive less light and the point-25 potential falls. With less tube-C current, point 28 rises so that tube A and thyratron 1 pass current; meanwhile, tube CC passes more current to lower the point-29 potential, turning off tube AA and thyratron 2. The current through the motor-M armature and tube 1 makes motor M steer the drive wheel farther off the black line.

19–4. Fast Correction for a Short Time. When a driver moves a car's steering wheel to quickly change the car direction, he may give the wheel a quick turn but then will bring the wheel back to its central position. In the same way, when the amount of light changes at the phototubes in Fig. 19D, the point-25 potential

---

11R and 4SX in series; the resulting a-c voltage, between tube-1 cathode 3 (at $S1T$ center tap) and the tube-1 grid (at point 45), fires tube 1 very late in its half cycle if 4SX receives very little d.c. through tube A. When the tube-A current increases, the voltage decreases across the a-c winding of 4SX, so that tube 1 is fired earlier in its half cycle; this increases the average voltage that tube 1 applies to motor M, raising the motor speed.

changes quickly to cause immediate steering action by motor M.*
However, as the motor-M speed rises, it makes tachometer T generate a voltage that offsets part of the phototube signal; this returns point 25 closer to normal, so as to slow motor M. Stability adjuster $3P$ may select the amount of this feed-back action desired from tachometer T.

Because of the sudden steering action that is necessary if the scanning head is to follow an abrupt change in black-line direction, this equipment may try to overshoot or overcorrect, causing the circuit to hunt. To help to stabilize this circuit in Fig. 19D, a notch network is included, as is explained in Sec. 18–11. Also, resistors 20R and 22R are included below tubes A and AA so as to decrease the time constant of the saturable reactors, to give greater stability as is mentioned in Sec. 18–13.

19–5. Manual Control. When it is desired to operate the tracing head by hand control, the contact of switch $S4$ is opened (at the lower right in Fig. 19D); this prevents the flow of any current through tubes A or AA. Instead, manual-control switch $S3$ now may be moved to right or left to operate the steering motor M. With $S3$ moved to touch at point 34, electrons flow from bottom line 9 through 18R, 2P, $S3$ and the d-c winding of 4SX; tube 1 passes current to turn the steering motor. If $S3$ connects instead to 35, 5SX receives d.c. so that tube 2 provides the current to turn the steering motor M in the reverse direction. When steered by hand, the speed of motor M is adjusted by $2P$.

When switch $S4$ provides manual operation, the movement of the steering head may let the light beam rest entirely on the black line, or entirely on the white background; in either extreme position, the electronic circuit becomes unbalanced and the various capacitors become charged by such unbalanced voltages. When $S4$ is returned to its position for automatic operation, these charged capacitors greatly delay the return of the tracing head to its on-line position. To prevent this delay, another contact of $S4$ is added between the anodes of tubes C and CC. When $S4$ is in the manual position, this added $S4$ contact ties together anodes 28 and 29, holding the electronic circuit nearly at the balance point. In this balanced condition, the circuit is ready to provide correct operation as soon as $S4$ is returned to the automatic position.

* Capacitor 6C provides speed-up or phase-lead action (as is described in Sec. 18–8) by causing the entire phototube signal at 27 to be applied to the tube-C grid for an instant, until 6C can change its charge through 5R.

CHAPTER 20

RUBBER CALENDER CONTROL

In the manufacture of rubber-coated fabric, groups of large calender rolls are driven by a single d-c motor, which receives its power from a motor-generator set, in the manner described in Sec. 16–1. This motor may need to run steadily at a speed that is only one-tenth of its top speed, or at any speed between; for threading the fabric or web through the rolls, the motor may run at only one-twentieth of its top speed. At any speed setting, the motor speed must remain constant; any desired change of speed must occur gradually, so as to hold constant tension in the fabric at all times.

20–1. Adjustable-voltage Calender Control (CR4815). To control the d-c calender motor, together with its motor-generator set and exciter, a large amount of magnetic control equipment is required, but is not discussed or included here. Often the calender is part of a series of rolls or auxiliaries, driven by separate motors whose speeds must be related at all times. Many electronic control panels are used in such a calender train; one of these is described below as an example.

20–2. Electronic Speed Regulator for Rubber Calender (CR 7513–E102). This control panel is shown in Fig. 20A; its diagram appears in Fig. 20B. The circuit operation in Fig. 20B is much like that of Fig. 16F, as described in Sec. 16–3. The speed of calender motor M is varied by weakening its own field (by thyratrons 3 and 4, shown at the bottom of Fig. 20B), or by changing its armature voltage as supplied by generator G (controlled through thyratrons 1 and 2).

Both the motor field and the generator field are controlled from one speed dial, at the upper right. With its slider 59 touching at point 53, for low speed, the potential of tube-$A1$ cathode 50 also is low; in brief, there is little tube-$C2$ current, but more current flows in tube $A2$, less in tube $B1$ (acting as a long-tailed pair). Very little current flows in tube $D1$ or GSX, so generator G supplies low armature voltage; full current flows in tube $D2$ and MSX, so

Fig. 20A. Electronic speed regulator for rubber calender (CR7513–E102).

motor M has full field current and turns at low speed. When the speed dial is in the high-speed position, slider 59 and cathode 50 are at high potential; greater current flows in tubes $C2$ and $B1$, less in tube $A2$. Tube $D1$ and GSX have large current flow, so generator G supplies full armature voltage to motor M; tube $D2$ and MSX have smaller current, so the motor field is weakened, letting motor M turn at high speed. Motor M drives the rubber calender and also drives tachometer generator T, which produces a d-c voltage that provides a true signal of motor-M speed.

For threading the calender, at very low speed, contact C–TH is closed (at the right in Fig. 20B, by a push button not shown); the circuit then responds to the potential at the slider of 8P, since any higher potential at slider 59 is absorbed across 50R.

Let us study these circuits in detail.

20–3. Conditions before the Motor Is Started. Most of the circuits in Fig. 20B operate from the d-c voltages between top point 16 and bottom points 12 and 13. From the a-c control lines, a constant-voltage transformer (not shown) supplies voltage to transformer 3T; from the $S3T$ winding at upper left, current passes through the selenium rectifiers 1Rec–4Rec and reactor 1X so that d-c voltage appears between 16 and 12. The voltage-regulator tubes G and H hold line 14 steadily at 105 volts above 13, and hold 15 at 105 volts above 14 (see Sec. 2–4, footnote). Top line 16 is about 95 volts above 15 because of the voltage drop across buffer resistor 18R.

Until a button is pushed, to start the calender motor, various relays are not picked up; the C–OR contacts remain closed (at the right in Fig. 20B), shorting capacitor 21C. Since point 52 is thus held at the potential of center line 14, the grid of tube $A1$ is also at this potential; enough electrons flow through 46R and tube $A1$ so that its cathode 50 also is near to point 14. Between cathode 50 and tachometer-T terminal 48 is a voltage divider 44R–45R–2P that controls the grid potential of tube $C2$. Since motor M is not turning the tachometer, T generates no voltage, so tube-$C2$ grid 47 is at the potential of center line 14. The cathode of tube $C2$ is raised above 14 by bias adjuster 6P so that little current flows through 40R or tube $C2$; anode 43 is near point-15 potential, raising point 42 and the tube-$A2$ grid.

The potential at 42 controls both tubes $A2$ and $B1$, which act as a long-tailed pair.[12–8] Electrons flow from 13 through 7P and 35R to point 37, then through 34R, tube $A2$ and 33R to 15; thus the voltage across 7P and 35R holds point 37 and the tube-$B1$

cathode at higher potential than the tube-$B1$ grid, "tied" through $30R$ to point 14. Because of the high potential at point 42, current flows through tube $A2$ but not through tube $B1$.

Because of the current flowing through $33R$, point 27 is lowered so that grid 30 of tube $D1$ is below point-14 potential; very few electrons flow through $8R$, tube $D1$, $26R$ and GSX. (Since grid 30 is below the potential of the $3P$ slider, no electrons flow through tube $C1$.) In the phase-shifting circuit[13-6] at the lower right, the GSX a-c winding has high inductance (because of the low current in the GSX d-c winding) so that the $S2T$ voltage fails to fire thyratrons 1 and 2; there is no field current, so generator G supplies no voltage.

Meanwhile, with little current through $31R$ and tube $B1$, anode 28 is at high potential and raises the potential of grid 24; electrons flow through $9R$, tube $D2$, $21R$ and MSX. Therefore, at the lower left, the MSX a-c winding has less inductance so that thyratrons 3 and 4 are fired early in their half cycles (similar to the action described in Sec. 14-6 and Fig. 14G) and supply full current to the field of motor M. This field current flows also through $9R$, producing a voltage that raises the tube-$D2$ cathode 23 above 14; this "feed-back" voltage across $9R$ helps to stabilize or regulate the action of tube $D2$ (see Sec. 18-13). The maximum field current may be adjusted by $4P$, acting through tube $B2$; if $4P$ is turned downward toward 14, electrons flow from $4P$ through tube $B2$ and $22R$ so as to lower grid 24, holding it close to the $4P$-slider potential.*

20-4. Operation at Low Speed. Here let us connect together points 52 and 57 (at the right in Fig. 20B); this removes the timed-acceleration circuit, to be explained later. The start button now operates various relays (not shown), opening the C–OR contacts so that point 52 is at the potential of slider 59 of the speed dial.

As slider 59 is turned to higher potential, for higher speed, point 52 and the tube-$A1$ grid rise also. At once cathode 50 follows the grid upward, raising point 47 and the grid of tube $C2$. Current increases through $40R$ and tube $C2$ so that its anode 43 drops, lowering point 42 and the grid of tube $A2$.† As current

---

* Tubes $B2$, $C1$, E and F act as simple diode rectifiers. All vacuum tubes in Fig. 20B are industrial-type twin triodes (5692); the use of this single type for all purposes simplifies tube replacement.

† Resistors $41R$ and $43R$ act as a voltage divider so that, as point 47 rises, part of the corresponding drop at point 42 feeds back to the tube-$C2$ grid, tending to degenerate or stabilize this tube circuit.

314 ELECTRONIC MOTOR AND WELDER CONTROLS

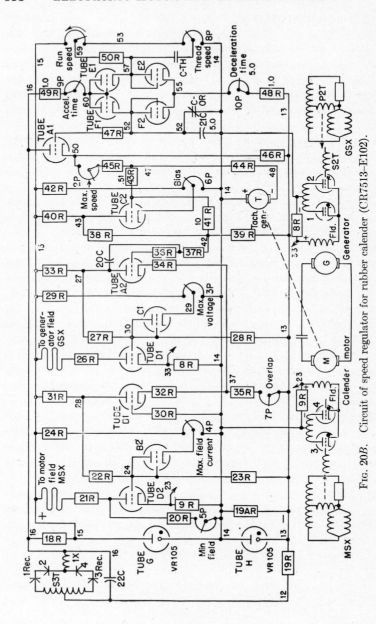

Fig. 20B. Circuit of speed regulator for rubber calender (CR7513–E102).

now decreases in 33R and tube A2, anode 27 rises; grid 30 of tube D1 likewise rises, so that more d.c. flows in reactor GSX. This decreases the inductance of the a-c winding of GSX (at lower right) so that the S2T voltage advances in phase, firing thyratrons 1 and 2 earlier and increasing the field current of generator G. As G supplies greater voltage to the armature of motor M, the motor speed rises and tachometer T generates more voltage.

The increasing speed voltage of T drives point 48 more negative, thereby lowering point 47 and the grid of tube C2. We see that the speed dial and tube A1 have set a voltage 50-to-14 that "asks for" the desired motor-M speed. The other circuits then raise or lower the motor speed until T generates just enough voltage 48-to-14 to match or offset the 50-to-14 voltage; any slight unbalance between these two voltages is sensed at the tube-C2 grid so as to correct further the speed of motor M.*

20-5. Medium Speed; Overlap. When slider 59 is turned to mid-position,† GSX receives its greatest current through tube D1, so that generator G supplies full voltage to the motor-M armature. This maximum voltage may be adjusted by 3P. When the tube-D1 grid 30 is raised to a potential equal to that at the 3P slider 29, electrons flow from 3P through tube C1 and 27R so as to connect grid 30 to slider 29, even though point 27 may be raised higher.

Although the generator-G voltage thus reaches its full value at mid-position of the speed dial, point 50 may be raised to still higher potential so as to weaken the motor-M field, to operate the motor above base speed. As this rise at 50 correspondingly lowers anode 43 and grid 42, current decreases in tube A2, 35R and 7P, lowering the potential at point 37. Since this lowers also the cathode of tube B1 (while the tube-B1 grid remains at point-14 potential) a point is reached where the current increases in 31R and tube B1. If all the resistance of 7P is in the circuit, point 37 still is more positive than 14 when the tube-A2 current already has decreased enough to let tube D1 cause full generator voltage. However, if 7P is turned clockwise to cause greater overlap, the

---

* Since the T voltage is a true indication of motor-M speed, there is little need for the IR-drop-compensation circuit such as is shown in Figs. 16F and 16H.

† Here the resistance of 8P is neglected and it is assumed that motor M has a top speed of twice its base speed. If a motor is used whose top speed is three times its base speed, the described results occur when slider 59 is turned upward one-third of its range.

same tube-$A2$ current causes less voltage across $7P$, so point 37 drops below 14 at a lower speed-dial position than before. In this way the weakening of the field current overlaps or starts before the point where full generator field is obtained.

20–6. Operation at High Speed. As the speed-dial slider 59 is turned near to its high-speed position, points 52 and 50 rise further; points 43 and 42 drop lower. As the tube-$A2$ current continues to decrease, the potential of point 37 drops steadily, increasing the current flowing in $31R$ and tube $B1$. This lowers anode 28 and the grid-24 potential of tube $D2$; tube $B2$ disconnects grid 24 from the $4P$ slider. Current decreases through tube $D2$ and MSX. At the lower left, the MSX a-c winding has greater inductance, so that thyratrons 3 and 4 are fired later in their half cycles of anode voltage; the current decreases in the field of motor M, which now runs above base speed.

At the top-speed setting, very little current flows through $D2$. However, even if tube $D2$ should fail, MSX receives current through $20R$ and $5P$, which are connected around tube $D2$. In this way $5P$ may adjust the minimum field current of motor M, and thereby may set the top motor speed.

20–7. Timed Speed-up and Slow-down. As long as points 52 and 57 are jumpered together (as is suggested in Sec. 20–4), the speed of motor M may change suddenly to bring the calender up to speed, when the speed dial is turned quickly. However, a main requirement of calender drive is that the motor speed must change slowly and smoothly, except during stopping; to see how the nearby circuits provide this feature, let us disconnect 52 from 57.

Briefly, during speed-up or acceleration, tubes $F1$ and $E2$ conduct; the current through $F1$ increases the $21C$ voltage so that the motor speed rises at the rate set by $9P$. During slow-down or deceleration, tubes $E1$ and $F2$ conduct; the current through $F2$ decreases the $21C$ voltage so that the motor slows down at the rate set by $10P$.

Suppose that the speed dial is preset for medium speed, so that slider 59 or point 57 is perhaps 50 volts above center line 14. Tubes $E1$, $E2$, $F1$ and $F2$ act as simple rectifiers, to switch or "take over" the control as follows. When the start button is closed and the C–OR contacts open across $21C$, point 52 is at point-14 potential and is rising slowly as capacitor $21C$ becomes charged to higher voltage; these charging electrons flow from $21C$ through tube $F1$, $9P$ and $49R$ to point 16. Acceleration-time adjuster $9P$ controls the amount of this charging current; when $9P$ has more resistance, point 52 rises less rapidly, bringing motor M up to speed more slowly. So long as point 60 still is more negative than point 57, there is no current through tube $E1$. Meanwhile, electrons flow from 13 through $48R$, $10P$, tube $E2$ and $50R$ to slider 59; cathode 55 is nearly 50 volts above 14, so there is no current through tube $F2$.

Since capacitor $21C$ is being charged through tube $F1$ to the voltage 16-to-14, point 52 rises until it reaches the preset potential of slider 59; then electrons flow from 59 through $50R$, tube $E1$, $9P$ and $49R$ so as to hold point 60 at slider-59 potential. Motor M rises to this preset speed and remains there. If slider 59 then is turned to yet higher speed potential, $21C$ is charged further through tube $F1$, to bring motor M slowly to the higher speed.

When the speed-dial slider 59 is turned downward to select a lower speed, the new potential at 57 is below that of point 52. Electrons still flow through tube $E1$, connecting point 60 to 57; anode 60 of tube $F1$ now is below point 52 so there is no tube-$F1$ current. Similarly, the lowering of point 57 stops the current flow through tube $E2$. Capacitor $21C$ now discharges; its terminal 52 slowly becomes more negative as electrons flow from 13 through $48R$, $10P$ and tube $F2$ into $21C$. The motor-M speed decreases slowly, at a rate set by $10P$, until point 52 has fallen to the potential at slider 59; then electrons flow through tube $E2$ to connect 55 to 57 and to hold point 52 at the potential of slider 59.

When any stop button is pushed, the motor is stopped as quickly as possible. The C–OR contact discharges $21C$ so that point 52 drops quickly to point-14 potential, thereby strengthening the motor-M field but turning off thyratrons 1 and 2. Motor M is stopped quickly because it pumps power back into G. The armature voltage of generator G is rapidly reduced to zero because the G field is connected to a Thyrite resistor (not shown), which is selected so as to limit the pump-back current to the highest safe amount.

CHAPTER 21

MULTICOLOR PRINTING

When a printer tries to place two or more colors accurately (one on top of the other or side by side) on a continuous strip of paper passing through a unit-type multicolor press, he may use a photoelectric color-register equipment to control the position of each printing cylinder.* To print a good color picture, the portions made by one color cylinder must match or line up with the portion printed in another color by the preceding cylinder, usually within an accuracy of $\frac{1}{200}$ inch. After receiving one color, the paper web may travel 15 to 40 feet (to permit the ink to dry) before it reaches the second color cylinder; here the second color must line up with the first regardless of changes in the stretch, tension or moisture during the long travel distance.

21-1. Color-register Control (CR7505–W116). Several color cylinders of a rotogravure press are shown in Fig. 21A, with the paper web passing between them. While the first color pattern is being printed (usually in yellow), that cylinder prints also a line of small register marks near the edge of the web; these marks are spaced accurately, twelve per revolution of the cylinder. When this printed paper nears the second color cylinder, the yellow marks are "watched" by a phototube in the web scanner W; the electric pulses from this phototube act through an electronic circuit so that a correcting motor A advances or delays the second cylinder to keep it exactly in line or in register. Connected to the shaft of the second cylinder is a scanner C shaped as a drum; it has twelve narrow slits evenly spaced around its rim. From inside this cylinder scanner, a light source D sends a beam through the slits to reach a phototube E; this phototube produces 12 electric pulses for each revolution of the second cylinder. If each pulse from web scanner W occurs just at the end of the corresponding pulse from phototube E in the cylinder scanner, the

---

* The control system described here applies to rotogravure presses. For web-fed offset presses, this system is modified but the electronic circuits remain the same.

second cylinder is correctly in line to print its color to match the first color.* If the W pulse arrives earlier during the E pulse, motor A corrects the cylinder position by moving it ahead, by means of the differential gear B; when the W pulse arrives late, motor A turns in the reverse direction to delay the second cylinder. A duplicate system positions the third cylinder so that it applies its color in the correct location, compared to the same yellow register marks.

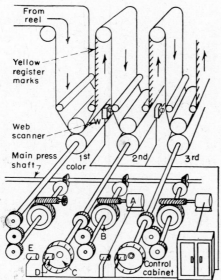

Fig. 21A. Portion of rotogravure press with color-register control.

Such color-register equipment is shown in Fig. 21B, for the control of several color cylinders; the circuit diagram appears in Figs. 21C and 21D. At the upper right in Fig. 21B, field-rectifier tubes E and F and control rectifiers C and D supply the voltages for the four duplicate panels that control four separate motors and color cylinders; one such motor panel is shown in the lower portion of Fig. 21B and is described in Sec. 21–2.

At the upper left in Fig. 21C, rectifier tubes E and F supply d.c.

* The use of 12 slits and 12 yellow marks per cylinder revolution gives greater accuracy or closer register than the use of fewer slits and marks. First the cylinder must be brought into register by manual control; then phototube E is moved to bring a drum slit into line with a register mark. Thereafter, by automatic operation, the electronic circuits keep this pair in line.

Fig. 21B. Photoelectric color-register control for four colors (CR7505-W116). Single color panel below.

MULTICOLOR PRINTING

to the fields of the correction motors. Similar tubes C and D produce 350 volts d.c. between points 5 and 4 for the operation of tubes 1 to 9 in each motor-control panel; this 5-4 voltage extends also to the speed-relay and web-break-relay circuits at lower right in Fig. 21D. Part of this 5-4 d-c supply is filtered also by 25C, 76R and 9C so that 300 volts d.c. is applied, between points 92 and 4, to the scanner circuits in Fig. 21D.

21-2. Outline of Circuit Operation. When the second cylinder is correctly in line with the yellow register marks (printed on the paper web by the preceding cylinder), a slit in the cylinder drum lets light reach phototube 11 (lower left in Fig. 21D) for an instant just before phototube 13 sees the yellow mark. The tube-11 signal passes through tubes 12, 1 and 2 to the No. 3 grid of tubes 4 and 5; the tube-13 signal passes through tubes 14, 15, 16 and 3 to the No. 1 grid of tubes 4 and 5. Tubes 4 and 5 send equal signals through both halves of tubes 6, 7 and 8 so that the current in saturable reactor 1SR is equal to the 2SR current; thyratrons A and B pass small but equal currents so that motor M does not turn in either direction.

If phototube 13 sees the yellow mark slightly before the cylinder-drum slit cuts off the light flash on tube 11, tube 4 delivers (through 16C and tube 6A to 23C) a stronger impulse than tube 5 sends through 17C and tube 6B to 22C; current increases through tubes 7A, 8B and 9A and reactor 1SR, so that thyratron A sends more current through motor M, turning M and the differential gear so as to move ahead or advance the second cylinder, thereby bringing it back into step with the yellow mark. Similarly, if phototube 13 sees its yellow mark after the slit has flashed light onto tube 14, then tube 5 sends the stronger impulse to 22C, so that current increases in tubes 7B, 8A, 9B, 2SR and thyratron B; motor M turns in the opposite direction so as to delay the second cylinder into step.

Parts of Fig. 21C are like Fig. 14E, described in Sec. 14-6. In Fig. 21C, tubes 8 and 9 control 1SR and 2SR as do tubes 3 and 4 in Fig. 14E; thyratrons A and B are phase-shifted like thyratrons 8 and 9 in Fig. 14E. At the right, tube 17 is an electron-ray tube that gives signals as does tube 2 described in Sec. 14-6.

At the lower right in Fig. 21D, web-break relay 4CR remains picked up so long as yellow register marks continue to pass web scanner W; if the paper web breaks, 4CR drops out and its n-c contact (at the left in Fig. 21C) picks up relay XCR to stop the press. To prevent such action at low press speed (as during threading), relay 7CR remains picked up at low speeds; near 10 per cent speed the yellow marks pick up 4CR before the 7CR contacts close.

21-3. Scanner Circuits. The cylinder scanner and the web scanner W in Fig. 21A may need to be 100 feet away from the control cabinet. So that the signal pulses from the phototube may travel this distance, other tubes are added in each scanner to strengthen the signals. At the lower left in Fig. 21D, each flash of light at phototube 11 increases the current through 73R, raising the tube-12 grid. As the current increases through tube 12 and 74R, its cathode 6 also rises. Here tube 12 acts as a cathode follower; as the tiny phototube current raises point 38, tube 12 passes greater current that raises cathode 6. This sudden rise at 6 is carried through the cable and 3C to raise the control grid of tube 1 in the main panel.

Meanwhile, each passing yellow mark causes a sudden decrease in the light reaching phototube 13 in the web scanner. Before this light dip occurs, grid 28 is high so that electrons flow from 4 through 61R, tube 14A and 58R to 92; the resulting voltage across 61R holds the tube-14B cathode above its grid (which is at point-4 potential) so that little current flows in tube 14B and 59R. Capacitors 24C and 32C are charged to the voltage between points 21 and 4. Since grid 23 is at point-4 potential, enough electrons may flow through 66R, tube 15, 64R and 62R so as to raise the tube-15 cathode slightly, to the point of balance. Similarly, 28C is charged to the 30-to-4 voltage and grid 31 is at point-4 potential; electrons flow through 68R, tube 16A, 71R and 72R. Grid 34 of tube 16B is about 20 volts above point 4, because of the voltage divider 70R-90R-72R; tube-16B cathode 7 is about 22 volts above point 4 but the control grid of tube 3 is at its cathode-4 potential.

As each yellow register mark passes, the grid-28 potential drops suddenly; the decreased current in tube 14A and 61R lowers the cathode potential of tube 14B, increasing the current in tube 14B and 59R. As anode 22 drops, grid 23 is driven negative so that current decreases in tube 15 and 64R; this raises anode 30 and, through 28C, raises grid 31 so that current increases through 16A and 71R. Through 29C, this lowers grid 34 so that the current decreases through tube 16B and 69R. Cathode 7 falls to lower potential; carried through the cable and 4C, the potential drops at $12P$ and the tube-3 control grid. We see that each yellow mark passing scanner W suddenly turns off tube 3; each flash of light through the cylinder scanner suddenly turns on tube 1.

21–4. Circuit Conditions without Signals. Before tubes 1 and 3 receive these signals from the scanners, no current flows in tubes 4, 5 or 6. The cathodes of tubes 4 and 5 are held at $+170$ volts at point 26 of the voltage divider $18R$–$20R$–$15R$–$25R$–$26R$–$24R$–$21R$; meanwhile, their control grids connect through $9R$ and $10R$ to a slider of lower potential, set to cut off all tube current. The cathode-26 potential also holds positive the grid of tube $2A$; enough current flows through $4R$ and $17R$ so that anode 28 is about 77 volts above point 4. This voltage appears across $2C$ (since electrons flow from cathode to grid of tube $2B$, and through $7R$ to charge $2C$); the tube-$2B$ grid voltage is near zero so that anode 29 is about 90 volts above 4. The No. 3 grids of tubes 4 and 5 are seen to be far more negative than their cathodes.* Since no tube current flows, anodes 44 and 43 are at the high potential of point 5; capacitors $16C$ and $17C$ are charged through $44R$ and $45R$ to the voltage between points 5 and 34.

Meanwhile the grid of tube $7B$ and the tube-$6B$ anode are at the potential of point 37 on the divider. The tube-$7A$ grid and the tube-$6A$ anode 57 are adjusted by $16P$ to near the point-37 potential to balance the currents in thyratrons A and B so that the correction motor does not turn. No current flows in tube 6; enough current flows through tube 7 and $33R$ so that cathode 53 is near the point-37 potential.

The tube-7 anodes are held about 170 volts above point 4 so that grids 71 and 72 of tube 8 also are at this potential. In Fig. $21C$, enough electrons flow through $53R$ and tube 8 so that cathode 62 is about 175 volts above 4. The equal currents in tubes $8A$ and $8B$ cause equal voltages across $78R$ and $79R$ and across $46R$ and $51R$. Anode 10 is about 6 volts below point 47, so about 0.2 ma flows from 10 through the $5CR$ contact, $48R$, $1SR$ and tube $9A$ to 47; similarly, 0.2 ma flows through $2SR$ and tube $9B$. As is shown in (b) of Fig. $14G$, these small equal currents in $1SR$ and $2SR$ cause thyratrons A and B to fire late in their half waves of

---

* Tube 4 or 5 has five grids and is used as a mixer tube to respond to several signals. No current flows in the tube unless the turn-on signal at No. 1 grid (nearest to cathode) occurs at the same time as the turn-on signal at No. 3 grid. The No. 2 and No. 4 grids prevent interaction between the two turn-on signals.

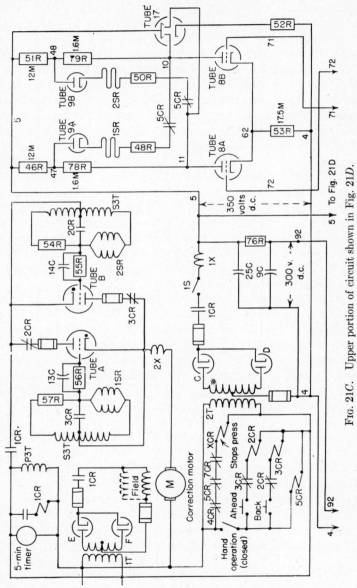

Fig. 21C. Upper portion of circuit shown in Fig. 21D.

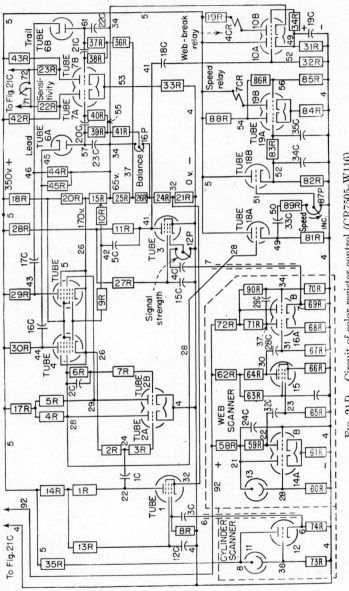

Fig. 21D. Circuit of color-register control (CR7505–W116).

anode voltage, to supply small equal and opposite pulses of voltage to the M armature; M does not turn.

21-5. Leading and Trailing Zones. As each turn-on signal is received at the grid of tube 1, its anode 22 drops suddenly, then returns quickly upward as the cylinder-scanner slit cuts off the light beam from phototube 8. Acting through capacitor $1C$ and $3R$ this drop at 22 forces negative the grid of tube $2A$ so that its anode 28 rises as shown at A in Fig. $21E$; an instant later, at B, the light flash ends and 28 returns to its starting value. This momentary rise at 28 raises also the No. 3 grid of tube 4; during this leading zone, current may flow through tube 4 but only when its No. 1 grid also is positive.

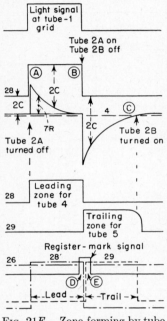

FIG. 21E. Zone forming by tube 2 in Fig. 21D.

When anode 28 rises at A, more electrons flow to the tube-$2B$ grid and $7R$ to charge $2C$ to a greater voltage. When 28 drops at B, capacitor $2C$ cannot lose this extra voltage at once, so $2C$ forces the tube-$2B$ grid below cutoff; at once anode 29 of tube $2B$ rises, thereby letting current flow in tube 5 during this trailing zone, but only when its No. 1 grid also is positive. As $2C$ now discharges through $6R$, the tube-$2B$ grid rises until, at C, current again flows through tube $2B$ and $5R$, lowering grid 29 of tube 5 to end the trailing zone.

Lower in Fig. $21E$, circuit conditions are shown when the second printing cylinder is exactly in register with the yellow marks on the web. The signal from each register mark appears at the grids of tubes 4 and 5 just as the leading zone ends and the trailing zone begins. So, for less than $\frac{1}{1000}$ sec, there will be a pulse of tube-4 current at D, instantly followed by an equal pulse of tube-5 current at E. Before the yellow mark appears, electrons flow from 4 through tube 3, $11R$ and $28R$ to 5 so that anode 41 is about 20 volts above 4; capacitor $5C$ is charged to the voltage between 41 and 42, the tube-5 grid. As each yellow mark passes, part of its signal (selected by 12P) is used to turn off tube 3, so its anode 41 rises suddenly; through 5C this pulse raises the No. 1 grids of tubes 4 and 5.

21-6. Shaping the Signal Pulses. When current flows through 30R and tube 4 for the brief instant (D in Fig. 21E), anode 44 drops to lower potential; acting through 16C, this drop also drives cathode 45 of tube 6A lower than point 34. Electrons flow from 16C through tube 6A so as to charge 23C;* its terminal 57 becomes more negative and decreases the current through tube 7A. When point 57 first is lowered, its whole effect passes through 20C directly to the grid of tube 7A; an instant later, when 20C has charged through 40R to the increased voltage between 55 and 57, the tube-7A grid returns to the potential at 55. In this way 20C provides a forcing action[18-8] to give higher response for an instant.

Similarly, the sudden short dip at tube-5 anode 43 acts through 17C so as to lower cathode 46 of tube 6B. Each such dip lets electrons flow through tube 6B to charge 22C and to lower the point-61 potential. Through 21C, the grid of tube 7B is lowered, decreasing the current through tube 7B and 43R, thereby to raise the grid-72 potential of tube 8A.

So long as the second cylinder is exactly in register, the small equal pulses at D and at E place equal charges on 23C and 22C. (Although these pulses occur at different instants, the capacitors average their effects so that nearly smooth d-c potentials are applied to the grids of tube 7A and 7B, decreasing equally the current in each tube.)

21-7. Action to Correct the Web Position.† If the position of the second cylinder lags behind the yellow register marks, the leading and trailing zones appear later than before. The register-mark signal seems to appear earlier at the No. 1 grids of tubes 4 and 5, as is shown in Fig. 21F. Here the entire register-mark signal occurs during the leading zone, wherein tube 4 may pass current. Since none of the signal occurs during the trailing zone, there is no tube-5 current. With pulses of current through tubes 4

---

* Each pulse at cathode 45 is too short to be used directly for controlling the reactor and thyratron circuits in Fig. 21C. However, by letting each pulse add a charge into capacitor 23C (upper right in Fig. 21D), the increased voltage remains across this capacitor until it discharges through 39R, 41R and 16P.

† Although the correction motor changes the position of the cylinder to keep it in register, the press operator may use manual-control buttons labeled "Ahead" and "Back"; these terms refer to the relative movement of the web. To appear to move the web back, actually motor M moves the cylinder ahead.

and 6A but not through tubes 5 and 6B, the average voltage across 23C becomes larger than that across 22C; the tube-7A current decreases. Since tubes 7A and 7B act with 33R as a long-tailed pair,[12-8] the tube-7B current increases. The grid-71 potential rises to increase the tube-8B current; grid 72 is lowered, decreasing the tube-8A current in Fig. 21C.*

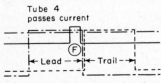

Fig. 21F. Correction signal to advance a lagging cylinder.

As is described in Sec. 14–6, greater current through tube 8B lowers its anode 10 and point 48. Decreased tube-8A current raises anode 11 and point 47. Since anode 11 may become more positive than point 48, the rectifier action of tube 9B stops the current flow through 2SR; the thyratron-B current stops. However, since anode 10 is lowered, greater current flows in tube 9A and 1SR; this increases the current through thyratron A, which turns motor M in the right direction to move ahead the second cylinder, to bring the zones back into line with the register-mark signal; the web appears to have been moved back.

Similarly, if the register-mark signal occurs too late, larger pulses of current pass through tube 5, less through tube 4. Capacitor 22C receives greater charge than 23C, and so the tube-7A current increases and the tube-7B current decreases. Tube 8B decreases the current through 1SR so that thyratron A is phased off. Greater current through tube 8A and 2SR now turns on thyratron B so that motor M turns to delay the second cylinder, thereby appearing to move the web ahead.

21–8. Control by Hand. At the lower left in Fig. 21C, a switch may be closed to permit hand control of the correction motor. The switch picks up relay 5CR whose contacts open, stopping all d.c. through 1SR and 2SR. To move ahead the position of the register marks, a button picks up 2CR. Above thyratron A, a 2CR contact opens to prevent any current through this tube; another 2CR contact connects the grid of thyratron B to its cathode so that full current flows through tube B to turn motor M so as to delay the position of the second cylinder. Similarly, another button may pick up 3CR to move the web backward; 3CR contacts turn on thyratron A so that motor M advances the cylinder.

* If the sensitivity adjuster is turned so as to decrease its resistance, more of the tube-7 output appears across 22R and 23R; there is less voltage change between 71 and 72 so the circuit sensitivity is reduced.

21–9. Web-break Relay. At the lower right in Fig. 21D is a relay 4CR that drops out and stops the press if the paper web breaks. As each yellow mark is "seen" at the web scanner, the tube-3 anode 41 rises; acting through 18C, this positive pulse raises the tube-10A grid. Each pulse of current through tube 10A produces a voltage across 31R that charges 19C and raises the tube-10B grid. Meanwhile, the tube-10B cathode is held at point 34, 65 volts above point 4. Above 10 per cent press speed, the tube-10A pulses occur so often that the 19C voltage holds the tube-10B grid close to cathode-34 potential; relay 4CR is picked up. If the paper web breaks, the yellow register marks no longer pass the web scanner. The positive pulses stop at point 41, so tube 10A cannot recharge 19C. Quickly 19C discharges through 31R; the tube-10B grid drops toward 4, so 4CR drops out, closing its contacts in Fig. 21C, stopping the press.

21–10. Speed Relay. At low press speed, or while threading the paper web through the press, the web-break relay should not stop the press. Therefore a speed relay 7CR is used in Fig. 21D; the circuits of tubes 18 and 19 keep 7CR picked up at low speed, so that the 7CR contacts remain open until the press reaches a speed sufficient to open the 4CR contacts mentioned above.

Each flash of light through the turning drum of the cylinder scanner causes the potential at tube-2A anode 28 to rise suddenly; these signals at point 28 are applied to the grid of tube 18A, causing pulses of voltage to appear across cathode resistor 82R. While the press is at standstill, no signals reach grid 28 of tube 18A, and its cathode 49 remains about 90 volts above point 4. However, 33C has charged to this voltage across 81R, so that point 50 and the tube-18B grid are at point-4 potential; tube 18B passes steady current and cathode 52 remains about 23 volts above 4, holding the tube-19A grid also at this potential. Grid 56 of tube 19B is held about 42 volts above 4, because of the voltage divider 85R–86R–88R. Enough current flows through tube 19B and 84R so as to pick up relay 7CR, opening the 7CR contacts.

At low press speed, each light flash holds 28 positive for a longer time than it does at higher press speed. To remove this variable effect, the time constant of 33C, 89R and 87P is made very short. At the start of each light flash (through a slit in the cylinder scanner), cathode 49 rises suddenly, and 33C causes a similar rise at 50 and grid 51 of tube 18B. Very quickly, before the light flash ends (even at top press speed), 33C has charged to the new voltage so that grid 51 has returned to point-4 potential; in this way each rising signal at grid 51 has the same length no matter how fast the press is running. Each rise at 51 lets more current flow through tube $18B$ and $82R$ for an instant, thereby increasing the charge in capacitor $34C$. The voltage across $34C$ increases as the press speed rises and is a true indication of press speed. When this $34C$ voltage (filtered by $83R$ and $35C$) increases to about 30 volts, current flows through $88R$ and tube $19A$ so as to lower the anode-54 potential. This causes a similar drop at grid 56, decreasing the tube-$19B$ current so that relay $7CR$ drops out and closes its contacts. Above this press speed, relay $4CR$ can stop the press if the web breaks.

CHAPTER 22

PAPER-MACHINE CONTROL

Of the many types of electric drive for papermaking machines, only the multiple-generator system is described here. Each d-c motor that drives a part of the paper machine receives its power from a separate generator that is controlled by an electronic-amplidyne speed regulator.

22-1. The Paper Machine. The upper part of Fig. 22A shows the main parts of a papermaking machine, driven by the motors M. The prepared stock flows onto the couch (similar to a belt conveyor), where the wood fibers settle to form a mat, and most of the water is withdrawn by vacuum. The mat of fibers passes between press rolls and then around steam-heated dryer cylinders; the calender rolls press the paper to provide the desired surface finish, then the paper is wound at the reel. The whole machine may be 300 ft long and may produce a paper strip 300 in. wide; most sections of the machine are large and heavy, driven by motors of 50 to 400 hp each.

The flow of stock and the speed of the couch control the thickness of the paper produced. When making light papers, the machine may need to operate at 2000 ft per minute, perhaps ten times as fast as when it makes heavy papers. The speed of the entire machine is controlled from a single dial S, at the lower left in Fig. 22A. Since the mat of fibers has very little strength, each motor must drive its section of the machine at a speed that matches the speed of the preceding section. To allow for stretch of the paper web, a motor may need to drive its section slightly faster than the preceding section; this speed difference, or "draw," is adjusted by a rheostat D.

22-2. The Multiple-generator System. Figure 22A shows how each motor M receives power from its own d-c generator G; these generators form a motor-generator set, driven by a large a-c motor. Each motor M receives constant field current from exciter E; the speed of each motor is varied by changing the amount of armature voltage received from its generator G. Each motor M drives also

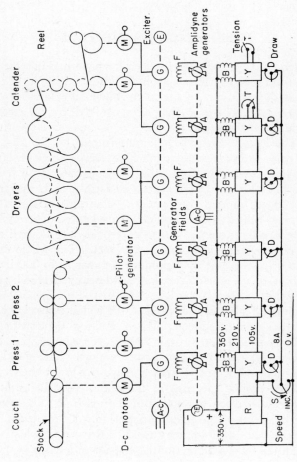

FIG. 22A. A paper machine with multiple-generator sectional motor drive.

a pilot generator *PG* (like the tachometer generator described in Sec. 17–6) that sends a motor-speed signal to the control circuit.

A smaller m-g set consists of amplidyne generators A, each supplying d.c. to the field F of its generator G; reference exciter *RE* supplies 350 volts d.c. for the amplidyne fields B, which are controlled by the electronic amplifiers Y (to be described in Sec. 22–5). Part of the 350 volts from *RE* is fed into the reference-voltage-supply unit R (to be described in Sec. 22–3), which produces regulated supplies of 210 volts and 105 volts d.c. for the amplifiers Y. Numerous magnetic-control circuits (buttons, relays, contactors, etc.) are not shown or described.

The control for one section of a paper machine is shown in Fig. 22B, including its amplifier Y.

Below each amplidyne generator A in Fig. 22A are shown two of its field windings B.* Each winding is controlled by a tube; current in the right-hand winding boosts the amplidyne-A output, but current in the left-hand winding opposes or bucks this output.

After the master dial S is set for the desired machine speed, each section motor M comes up to this speed when its "Run" button is operated. Electronic circuits in Y strengthen the "boost" field so that amplidyne A increases the field current of G; because the armature voltage increases, the motor-M speed rises until *PG* produces a speed voltage that matches the reference voltage

Fig. 22B. Control panel for one motor section shown in Fig. 22A.

* Each amplidyne has two additional fields, not shown. One controls the amplidyne generator at "breakaway," to provide enough torque while motor M is being started or while the "Slow" button is in control; the other field provides current limit so that motor M does not receive too much current from G while bringing its section up to the speed set by the master speed dial S.

selected by S (and the speed is influenced further by the draw rheostat D).

When the paper web leaves the dryers, the tension in the paper can be used to control the speed of the calender or reel motor. Through added circuits (described in Sec. 22–11), the tension rheostat T sets the maximum current that the motor may draw

Fig. 22C. Reference voltage supply (CR7590–AD102).

from G, and thereby makes the motor hold constant tension as the paper is calendered or reeled.

22–3. Reference Voltage Supply (CR7590–AD102). To hold constant paper-machine speed for long periods, the pilot-generator speed voltages must be compared with a voltage as set by S, which never changes. As is shown at the left in Fig. 22A, reference-voltage-supply panel R furnishes this constant voltage for master speed dial S and also for the electronic circuits within the Y amplifiers. This panel R is shown in Fig. 22C; its diagram appears in Fig. 22D.

The reference exciter RE furnishes 350 volts d.c. at the left in Fig. 22D; although this voltage is filtered by reactors and capacitor 1C, it may change owing to variations in main power supply or in RE temperature. However, tubes 1, 2, 3 and 4 (connected in parallel to act like a single tube) absorb these voltage changes so that constant voltage appears between lines 210 and 0. Similarly, tubes 7 and 8 act as another regulator to hold 105 volts between lines 105 and 0, so that the speed-reference voltage selected by dial S will remain constant.

Voltage-regulator tube 6 holds 75 volts across 2R and 3R so that the tube-5 cathode 14 remains about 55 volts above 0. The tube-5 control grid connects (through 16R) to slider 13 of the voltage divider 7R–2P–6R that is connected between 210 and 0. If for any reason the 210 potential rises, thereby raising the control-grid potential of tube 5, more electrons flow from cathode 14 through tube 5 and the large resistance 5R to point 9. This lowers the potential at point 11, connected to the grids of tubes 1 to 4; these tubes act together as a cathode follower (see Sec. 13–5), for their cathode potential 210 follows downward as the grid potential is lowered.* Or tubes 1 to 4 may be regarded as a variable resistance between points 350 and 210 and connected in series with the load circuits between 210 and 0; lowering grid 11 increases the resistance of the tubes so that more of the total voltage (350-to-0) appears across the tubes, less across the load (210-to-0).

Similarly, the output voltage 210-to-0 may be adjusted by 2P so that this reference panel R may supply a different number of Y amplifier units.† To increase this 210-to-0 voltage, the 2P slider 13 is turned to a lower potential, thereby decreasing the

---

* The combined resistance of all circuits connected between 210 and 0 acts as the common cathode resistor of tubes 1 to 4; to carry the amount of current flowing in these 210-to-0 circuits, four tubes are needed. As grid 11 is lowered, less current flows through these tubes and the cathode-circuit resistance, decreasing the voltage drop across these cathode circuits; the potential of cathode 210 is lowered.

† Adjuster 1P determines what portion of any variation in the voltage 350-to-0 shall reach the shield grid of tube 5. If 1P is set too high, a voltage rise at 350 may increase the tube-5 current and lower the 210 potential; if set too low, a rise at 350 may permit a rise at 210. Therefore 1P is factory-adjusted so that a change at 350 has no direct effect on tube 5.

current through tube 5 and $5R$; point 11 rises, increasing the current through tubes 1 to 4 so that point 210 rises.

22-4. The 105-volt Regulator. Although the 210-volt output is held constant by the circuits in Fig. 22D, tubes 7 and 8 are added as a second regulator to prevent any shift in the potential at mid-point 105 that might result from load changes in the output circuits.

Any small decrease in the 105-to-0 voltage lowers the potential of the tube-$7A$ cathode; meanwhile the tube-$7A$ grid is held at steady potential (at the $4P$ slider in voltage divider $18R$–$4P$–$17R$)

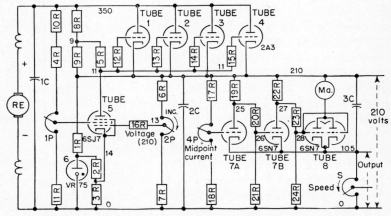

Fig. 22D. Circuit of reference voltage supply (CR7590–AD102).

by the regulated voltage 210-to-0. Thus current increases in $19R$ and tube $7A$, lowering anode 25 and the tube-$7B$ grid 26; current decreases in $22R$ and tube $7B$, raising anode 27 and the tube-8 grids 28. More current flows through tube 8 and its cathode load (consisting of speed dial S and other output circuits) so that the voltage 105-to-0 increases to its full amount.*

Tube 8 acts like a variable resistance in series with the 105-to-0 load resistance; so long as the tube-8 resistance exactly equals the load resistance, point 105 is held midway between 210 and 0. If the load resistance increases, point 105 rises, so that tube 7 decreases the tube-8 current; the resistance of tube 8 increases, lowering 105 to the desired mid-point potential.

* To operate tube 8 under the best conditions, $4P$ is adjusted until the tube-8 current is about 8 ma, as shown on the meter in its anode circuit.

The regulated voltages, supplied by the reference panel R (Fig. 22D) for all of the amplifier units Y, are shown at the left in Fig. 22E, to be studied next.

22-5. Amplidyne Preamplifier Unit (CR7513–K102G1). Each of the electronic amplifiers, shown as Y in Fig. 22A, has the circuit arrangement shown in Fig. 22E. This circuit controls one amplidyne generator (by its buck and boost fields at the upper right) so that one generator G supplies armature voltage for one motor M; the M speed is indicated by the voltage produced by its pilot generator PG, at the left. When the "Run" button is operated to let motor M drive its section at the speed set by the master dial S, relay $1CR$ closes its contact below tubes 3 and 4 but opens its other contacts in Fig. 22E. Tube 5 acts only while motor M comes up to the preset speed.

In brief, at running speed the voltage produced by PG offsets or balances against the S speed-reference voltage (105-to-S slider 8A) so that the grid of tube 1A is near to mid-point-105 potential. If for any reason the M speed and PG voltage increase, more current flows through tube 1A, less through tube 1B, more through tube 2A, less through tube 2B, so that tube 3 strengthens the buck field and tube 4 decreases the boost field; amplidyne A supplies less field current to generator G so that motor M receives less armature voltage and its speed decreases.

22-6. Circuit Conditions at Rest. All parts of the circuit of Fig. 22E receive power from the reference exciter (350-to-0 volts) or from the 210-volt and 105-volt regulated supplies described in Sec. 22–3.

Before the "Run" button picks up relay $1CR$, the $1CR$ contacts 15–16 are open so that no current can flow through tubes 3 and 4 or the amplidyne fields above them.* Other $1CR$ contacts (at the left in Fig. 22E) connect the pilot generator across 15AR and 16AR, to mid-point 105 and to point 93; since the PG voltage is zero or is very small, point 93 is held close to point-105 potential, so that capacitor 9C has very little charge. Since 68R has much less resistance than 10P or 53R, the tube-5B grid 92 also is close to point-105 potential; since tube 5B acts as a cathode follower,[13–5] its cathode 63 also is near 105 and (through tube 5A) holds point 64 close to point-105 potential. Point 64 is on the voltage divider 1P–61R–62R–63R–64R that controls the grid potential of tube

* See Sec. 22–2, footnote. The "Slow" button may be in control, turning M and PG at low speed.

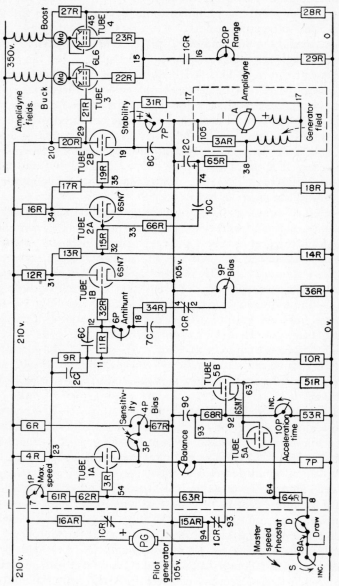

Fig. 22E. Circuit of amplidyne preamplifier unit (CR7513-K102G1).

$1A$; therefore, because of the closed $1CR$ contact 93–94, the tube-$1A$ grid is held near or above point 105 so that current flows through $4R$ and tube $1A$, and its anode 23 is at a potential not far above 105.

Connected between 23 and 0 is the voltage divider $9R$–$10R$ that holds the tube-$1B$ grid at a lower potential so that very little current flows in tube $1B$ and $12R$. (For a moment, let us omit the circuit through the $1CR$ contact 2–4 at the center of Fig. 22E.) Since anode 31 is at high potential, grid 32 lets current flow in $16R$ and tube $2A$ so that its anode 34 is at low potential, thereby holding grid 35 below 105 so that little current flows in $20R$ and tube $2B$. Since anode 29 is at high potential, near to 210, tube 3 is ready to pass full current to the buck field as soon as the $1CR$ contact 15–16 closes. Since tubes 3 and 4 act as a long-tailed pair,[12-8] the boost field then will receive little current. So, at the instant when the "Run" button picks up $1CR$, the voltage applied to motor M may not be large enough to start the motor turning. However, the voltage for starting motor M from rest may be adjusted by $9P$. Before starting, the $1CR$ contact 2–4 completes a circuit from the $9P$ slider, through $34R$, $6P$ and $11R$ to point 11; on this voltage divider, grid 12 of tube $1B$ may be held at higher potential (as set by $9P$) so that more current flows through tube $1B$ and tube $2B$. Capacitor $7C$ is charged by the voltage between mid-point 105 and point 18 on the divider. Therefore, for a few seconds after the "Run" button picks up $1CR$, the increased tube-$2B$ current lowers point 29; tube 3 decreases the buck-field current and tube 4 strengthens the boost field so that motor M receives greater armature voltage at breakaway. However, $7C$ soon discharges through $6P$ and $11R$, letting grid 12 return to the potential at point 11. Because of $7C$ and the $1CR$ contact 2–4, the setting of $9P$ can cause greater torque to let motor M break the machine away at the start; less torque then is needed to bring the machine up to speed.

22–7. Timed Acceleration. To control the rate of speed-up, so that motor M will bring its section up to speed gradually, tube 5, $9C$ and $10P$ are added in Fig. 22E. As is explained above, there is very little voltage across capacitor $9C$ just as the "Run" button picks up $1CR$. However, when the $1CR$ contact disconnects the $9C$ terminal 93 from 94, the voltage across this capacitor increases; electrons flow from 0 volts through $53R$, $10P$ and $68R$ to charge $9C$. Therefore the potential at tube-$5B$ grid 92 is lowered at a rate adjusted by $10P$; the cathode-63 potential follows grid 92 downward, letting point 64 fall slowly toward the potential at the speed dial S.

This gradual lowering of point 64 (on the divider $1P$–$61R$–$62R$–$63R$–$64R$) decreases the grid-54 potential of tube $1A$. Since current decreases in tube $1A$ (causing more current in tube $1B$, less in tube $2A$, more in tube $2B$ and less in tube 3, as is described above), more current flows in tube 4 and the boost field, so that the motor-M speed increases. Since M now drives PG at greater speed, PG produces greater voltage that raises its terminal 7 to higher potential; in turn, this raises grid 54 back toward the point of balance. As point 64 continues to fall, PG must turn faster so as to raise point 7; thus the grid-54 potential is permitted to decrease only a fraction of a volt, but this is enough to raise the speed of motor M and PG as desired.

As the charging of $9C$ continues, grid 92 and cathode 63 approach 0 potential. When 63 has dropped to the normal potential of point 64 (as determined by the voltage divider), tube $5A$ acts as a rectifier to disconnect 63 from 64; capacitor $9C$ has no further effect until motor M is started again.

22–8. Machine-speed Adjustment. After motor M has brought its section to the speed set by master dial S, the motor speed is controlled mainly by S.* As the S slider $8A$ moves toward 0, it raises the speed of all the motors driving sections of the paper machine. However, since the paper lengthens slightly as it passes through the machine, each section may need to run a little faster than the section preceding it. To provide this feature, a draw rheostat is used with each section. In Fig. 22E, draw rheostat D is shown as part of the voltage divider $1P$-to-$64R$ that controls the grid potential of tube $1A$. When each draw rheostat is in its middle position, all machine sections run at the same speed (as adjusted by $1P$).† When D is turned clockwise to decrease its resistance, this lowers the potential at all points along the divider, including the potential at grid 54; this raises the motor-M speed slightly so that this section handles the paper web as fast as it is received from the preceding section.

* The master speed dial S is a motor-driven rheostat, responding to a speed dial at the control desk. Although the operator may turn this dial quickly to a new speed setting, the motor drives the master speed dial S at a fixed rate, permitting only a gradual change in paper-machine speed.

† The draw rheostats of sections ahead of the first dryer are turned to decrease the draw, making these sections run at speeds less than the dryer speed. For sections after the dryers, the draw is increased.

Whether grid 54 of tube 1A is lowered by increasing the draw (by D) or by raising the machine speed (by S), the action is the same. The decrease of tube-1A current raises anode 23 so that points 11 and 12 rise also, increasing the current through 12R and tube 1B; since this lowers anode 31 and tube-2A grid 32, the tube-2A current decreases and its anode 34 rises. Since this raises grid 35 so that more current flows through 20R and tube 2B, point 29 is lowered; fewer electrons flow through 29R, 20P, 1CR contact, 22R, tube 3 and the buck field. Since the potential of point 15 is lowered but grid 45 of tube 4 remains at steady potential (on voltage divider 27R–28R), greater current flows in tube 4 and the boost field of the amplidyne. Amplidyne A provides greater current to the field of generator G, increasing the armature voltage of motor M so that its speed rises; PG produces greater voltage so that grid 54 is returned to its point of balance.

22–9. Circuit Details. At the right in Fig. 22E, range adjuster 20P selects the total amount of current (about 30 ma) flowing in beam-power tubes 3 and 4. If the resistance of 20P is decreased, the potential of point 15 is lowered; more current flows in tubes 3 and 4 to raise point 15 back to the point of balance.

Below tube 1A, adjuster 3P controls the sensitivity of the amplifier, as is explained in Sec. 16–7. When all the resistance of 3P is in the circuit (for lowest sensitivity), 2P may adjust the balance of current in tubes 3 and 4; similarly, 4P adjusts this balance when 3P is turned clockwise (for greatest sensitivity).

Below tube 2B, the voltage produced by the amplidyne-A armature is connected across 7P and 31R in series. If 7P is turned clockwise, increasing its resistance, a larger portion of the amplidyne voltage appears across 7P and capacitor 8C. As grid 35 rises, increasing the tube-2B current so that amplidyne A produces greater voltage, the voltage across 8C increases and raises cathode 19 to offset part of the rise at grid 35. This degenerative feedback is mentioned in Sec. 18–13; it makes amplidyne A respond faster to a speed signal at the tube-1A grid and helps to stabilize the circuit.

22–10. Antihunt Circuits. Other antihunt networks appear at the grids of tubes 1B and 2A. As is mentioned in Sec. 18–1, grid 12 of tube 1B receives speed-up signals through 2C and 9R, also through 6C and 11R. Slow-down signals received through 7C are adjusted by the resistance of 6P.

If the current in the generator field changes quickly, part of this change is fed through 10C to the grid of tube 2A. Capacitor 12C is charged to the voltage across resistor 3AR; if the generator-field current decreases so that there is less voltage across 3AR, 12C discharges through 65R to this lower voltage. Since this lowers the potential of point 74, the charge on 10C lowers briefly the potential at grid 33; through the resulting action of tubes 2A, 2B, 3 and 4, the boost-field current increases, raising the amplidyne-A voltage so as to restore the generator-field current to its

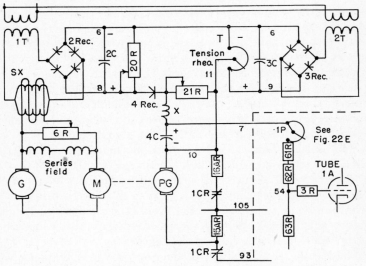

FIG. 22F. Circuit for tension control, added to Fig. 22E.

normal value for a short time. Similarly, a sudden change of amplidyne-A voltage changes the voltage across 8C, at the cathode of tube 2B, to help stabilize the system.

22–11. Tension Control. After leaving the dryers in Fig. 22A, the paper web may have enough strength to permit considerable tension or pull on the paper. Tension control may be added to the calender motor or the reel motor, so that the motor speed is regulated so as to hold constant motor current. Figure 22F shows this tension-control circuit added to the speed-control circuit of Fig. 22E; the voltage across capacitor 4C is added to the pilot-generator PG voltage so as to control tube 1A.

While motor M drives PG to produce a motor-speed signal, the motor-M armature current produces a small voltage across 6R that indicates the amount of motor load and tension; $6R$ is connected across the series and commutating fields of M and G. Part of this voltage across $6R$ forces direct current through saturable reactor SX; when the motor-M current and the $6R$ voltage increase, SX permits more voltage from transformer $1T$ to reach the metallic rectifier 2Rec, thereby producing more d-c voltage across $20R$ (filtered by $2C$).

Meanwhile, transformer $2T$ applies voltage to 3Rec so that a constant d-c voltage appears across tension-adjuster T (filtered by $3C$). So long as the voltage across $20R$ is less than the voltage 6-to-11 selected by the T slider, rectifier 4Rec prevents any current flow through $21R$, so there is no charge on $4C$; $4C$ has no effect on the motor-M speed.

When T is set for the desired motor-M current (or paper tension), the draw rheostat for this section is turned to raise the motor-M speed. As the motor pull or tension increases, more voltage appears across $6R$; SX becomes more saturated, and more voltage (rectified by 2Rec) appears across $20R$, so that point 8 becomes more positive than T-slider 11. Current now flows through 4Rec and $21R$; this produces voltage across $21R$ which, filtered further by X and $4C$, raises point 7 to a more positive potential than point 10. This signal, added by the voltage across $4C$, has the same effect on tube $1A$ (and the speed-control circuit of Fig. 22E) as though PG were being driven at too high speed; therefore the other tube circuits control amplidyne A so as to decrease the speed of motor M until it produces just the amount of tension selected by T.

INDEX

A

Acceleration control, 265, 339
Accuracy of follow-up, 183
Amplidyne, 210n., 278, 290, 333
Amplidyne preamplifier, 278, 337
Amplifier, capacitor-coupled, 207
    two-stage, 182n.
Anneal weld, 107
Antihunt circuits, 200, 253, 288, 341
Arc welding, 3n.
Armature control, 192, 234, 268
Armature feedback, 278
Automotive welding timer, 66
Averaging time, 13

B

Back-to-back tubes, 7, 10
Back voltage, 193, 227
Ballast tube, 210n.
Base speed of motor, 192
Beam power tube, 281n., 341
Bias, grid, 8n.
Bloodhound, electronic, 303
Bode diagram, 288, 290, 294
Brain (Thy-mo-trol control center), 231
Braking, dynamic, 246

C

Calender, rubber, 310–317
Calibration adjuster, 91n., 127n.
Capacitor-coupled amplifier, 205
Capacitors, series, 144
Cathode follower, 196, 273, 302, 322, 335, 337
Cathode resistors, 301, 309
Chill time, 110
Closed-cycle system, 184, 278, 289–302
Color-register control, 318
Constant current, 238, 285
Constant speed, 239
Contactor, ignitron, 5
Cool time, 24, 29, 48, 98, 141
Coordinator circuit, 161, 164
Copper-oxide rectifier, 7
Correction-time circuit, 208
Counter emf, 193, 227
Coupling circuit, 136
CR7500 GE controls:
  CR7503–A140, 75
    –A148, 146
    –A151, 83
    –A152, 95
    –A160, 109
    –A164, 112
    –B120, 114
    –C121, 124
    –D149, 107n.
    –D175, 10
    –D202, 100
    –D209, 175
    –DY2, 36
    –F180, 24
    –F220, 8
    –G119, 15
    –L102A8, 30
    –L102S15, 55
    –L127, 66
    –M101, 158
  CR7505–S118, 185
    –T102, 194
    –W116, 318
  CR7507–F170, 226
    –G270, 232
    –N103, 257
    –N115, 256
  CR7510–A102, 179
  CR7512–B100, 303
  CR7513–E101, 265
    –E102, 310
    –E118, 274
    –K102, 337

CR7500 GE controls:
   CR7515–S120, 212
     –S127, 209
     –W108, 220
     –W201, 203
   CR7590–AD102, 334
Current control, 238
Current limit, 239, 274, 283
Cutoff, 182n.
Cutoff-register control, 203, 220

D

D-c motor, control of, 192, 226–260
D-c supply, 191
Dead zone, 183, 190
Decade, 293n.
Decibel (db), 292
Degenerative feedback, 302, 341
Divider, voltage, 196
Dot inside tube circle, 10n.
Double-squeeze time, 72
Draw rheostat, 340
Drop-out time, 161n., 174
Dual-weld-time selector, 36
Duty cycle of ignitrons, 13
Dynamic braking, 246

E

Electron flow, 5n.
Electron-ray tube, 220, 222n.
Electronic amplidyne, 278–288
Electronic tracer, 303
Electronic variator, current limit, 274
   timed speed-up, 265
Error voltage, 182, 184
Exciter circuit, 168, 171

F

Feedback, armature, 278
   degenerative, 302, 341
   tachometer, 285
Field control, 192, 237, 266
Filter, 15n., 24n.
Flip-flop control, 15
Follow-up action, 182
Forcing, 201
Forge welding, 112

Frequency-changer welding control, 158
Full-field starting, 245

G

Gain, 292
Gas cutting torch, 303–309
General Electric controls (*see* CR7500)
Generator, amplidyne, 210n.
   field control of, 262, 268
Grid, shield, 27n.
   rectification, 39n.
Grid bias, 8n.

H

Half-wave Thy-mo-trol drive, 226
Heat control, phase-shift, 10
   of welder, 9
Heat time, 22, 29
Heat to make a weld, 3n.
High-voltage welding control, 146
Hold time, 21, 63
Hunting, problem of, 289
Hunting action, 183, 200, 215

I

Ignitor, 5, 6
Ignitron, rating of, 13, 146n.
Ignitron contactor, 5
Indicator tube, 220, 222n.
Inductance of saturable reactor, 199, 216
Inductive hangover, 58, 59
Industrial-type tube, 313n.
Initial squeeze time, 72
Inversion control, 161n., 174, 175
Inverter action, 18, 249
IR-drop compensation, 240, 272, 283, 315n.

L

Ladder, 196
Lag network, 295
Larger loads, ignitron controls for, 14
Lead network, 298
Leading-tube–trailing-tube circuit, 75
Light relay (*see* Photoelectric controls; CR7505)

INDEX

Line-follower action, 308
Load ratings of ignitron tubes, 13, 14, 146n.
Long-tailed pair, 187
Loop control, 194
Lower frequency for welding, 156

M

Mercury cathode, 5
M-g set for speed control, 261
Mixer tube, 222, 323n.
Motor, d-c, 192
Motor controls, 177–343
Motor-generator control of speed, 261–277
Multicolor printing, 318–330
Multiple-generator system, 331
Multitransformer welding machines, 155

N

Negative-control tube, 10n.
NEMA 1A weld timer, 7n.
    3B, 20, 30, 37, 55, 66
    5B, 20, 24, 42, 124
    7B, 84, 130
    9B, 95
    N2, 30, 49, 55, 66
    N3, 42
    S2H, 85, 129
    S3H, 95, 137
Network, 295n.
Nonbeat timer, 20n.
Nonrepeat welding, 63
Normally-closed contacts, 33
Notch network, 288, 298, 300, 309
Number-of-spots circuit, 127
Nyquist approach, 290n.

O

Off time, 22, 30
One-way cutoff control, 203
Overlap control, 260, 266n., 270, 315
Overvoltage control, 238

P

Paper machine, 331
Peaking transformer, 136n.

Pentode, 196n.
Per cent duty of ignitrons, 13
Phase-lag network, 295
Phase-lead network, 298
Phase-shift heat control, 10, 123
Phase shifting, 197, 216
Photoelectric loop control, 194
Photoelectric register control, 203–225
Photoelectric tracer, 303–309
Pilot generator, 333
Position controls, 179–190
Positive-control tube, 45n.
Power-factor angle, 12
Printing control, 318–330
Pulsation welding, 22, 29, 48
Pulsation-weld combination, 95, 124, 137
Pump back, 263, 275, 284, 317

Q

Quadrature circuit, 291, 293
Quick slow-down, 253

R

Radians, 292n.
RC time constant, 9n.
Reactive ohms, 229n.
Reactor, saturable, 197, 199
Rectifier, 24n.
Reference voltage, 279, 285, 334
Register controls, 203–225, 318–330
Regulator, 336
Regulator tube, 24n.
Repeat welding, 30, 33, 42, 48, 64
Resistance load, thyratrons with, 57
Resistance welding, 3
    (*See also* CR7503)
Rubber-calender control, 310–317

S

Saturable reactor, 197, 199
    web correction by, 215
Scanner, 322
Seam welding, 114
Seam-welding control, 114, 124, 139
Selsyn, 184, 257
Sensitivity control, 270, 341
Sequence weld timer, 20–73

Series capacitors, 144
Shaping, of motor field, 267
  of signal pulse, 327
Shield grid, 27n.
Shunt d-c motor, 191
Side-register control, 185, 209, 212
Signal gain, 291
Slope, unit, 293
Slope control, 100
Slow-down, by dynamic braking, 246
  quick, 253
Solenoid control of motor speed, 256
Speed-drop adjuster, 240
Speed limiter, 287
Speed relay, 329
Speed variator, 261–277
Spot-weld combination, 83, 124, 129, 146
Squeeze time, 21, 62
Stability, 288, 289–302, 309
Sudden light changes, 205
Synchronous combinations, 83–99, 106, 109, 112, 124, 129–143
Synchronous control, 75, 83–99
Synchronous timing, 74
System, closed-cycle, 184, 278, 289–302
  multiple-generator, 331

T

Tachometer feedback, 285
Tachometer generator, 285, 307, 309, 312, 333
Take-over tube, 259, 272, 274, 316
Taylor Winfield N3 timer, 42
Temper time, 110
Temper welding, 107, 109
Tension control, 256, 310–317, 342
Three-phase welding controls, 155–178

Thy-mo-trol drive, 226–260
Thyratron, 10n.
Thyratron with inductive load, 58
Time constant, 9n., 293
Timed-acceleration control, 265, 339
Timed slow-down, 272, 316
Timer, automatic weld, 20
Transient current, 74
Two-point register control, 185
Two-stage foot switch, 20n., 44, 49
Two starting switches, 36
Two-way cutoff control, 220

U

Unit slope, 293

V

Variator, speed, 261–277
Vector diagram, 11, 144, 199, 216, 229m.
Voltage-regulator tube, 24n.

W

Wave shapes, 247
  for inverter operation, 251
Web-break relay, 329
Web correction, 215
Web-register control, 203–225
Weld interval, 22
Weld time, 7, 21, 62
Welding-power circuit, 94
Weltronic 3B timer, 37
Westinghouse controls, N2 combination, 49
  seam-welding combination, 139
  sequence 7B timer, 130
  synchronous S2H combination, 129
  synchronous S3H combination, 137

Date Due			
MAR 18 '52			
DEC 16 '52			